T0275729

Friction Stir Casting Modification for Enhanced Structural Efficiency

Friction Stir Casting Modification for Enhanced Structural Efficiency

A Volume in the Friction Stir Welding and Processing Book Series

Saumyadeep Jana
Pacific Northwest National Laboratory, Richland, WA, USA

Rajiv S. Mishra
Department of Materials Science and Engineering,
University of North Texas, Denton TX, USA

Glenn J. Grant
Pacific Northwest National Laboratory, Richland, WA, USA

ELSEVIER

AMSTERDAM • BOSTON • HEIDELBERG • LONDON
NEW YORK • OXFORD • PARIS • SAN DIEGO
SAN FRANCISCO • SINGAPORE • SYDNEY • TOKYO

Butterworth-Heinemann is an imprint of Elsevier

Butterworth-Heinemann is an imprint of Elsevier
225 Wyman Street, Waltham, MA 02451, USA
The Boulevard, Langford Lane, Kidlington, Oxford OX5 1GB, UK

Notices
Knowledge and best practice in this field are constantly changing. As new research and
experience broaden our understanding, changes in research methods or professional practices,
may become necessary.

Practitioners and researchers must always rely on their own experience and knowledge in
evaluating and using any information or methods described herein. In using such information or
methods they should be mindful of their own safety and the safety of others, including parties for
whom they have a professional responsibility.

To the fullest extent of the law, neither the Publisher nor the authors, contributors, or editors,
assume any liability for any injury and/or damage to persons or property as a matter of products
liability, negligence or otherwise, or from any use or operation of any methods, products,
instructions, or ideas contained in the material herein.

ISBN: 978-0-12-803359-3

Library of Congress Cataloging-in-Publication Data
A catalog record for this book is available from the Library of Congress

British Library Cataloguing-in-Publication Data
A catalogue record for this book is available from the British Library

For Information on all Butterworth-Heinemann publications
visit our website at http://store.elsevier.com/

Working together
to grow libraries in
developing countries

www.elsevier.com • www.bookaid.org

CONTENTS

List of Figures vii
List of Tables xv
Preface to This Volume of the Friction Stir Welding and
Processing Book Series xvii

Chapter 1 Introduction 1
References 4

Chapter 2 Friction Stir Processing: An Introduction 5
Microstructure: Effect of Process Parameters 7
Recrystallization Mechanisms 13
Limitations in Refinement 24
References 26

Chapter 3 Mechanical Properties Enhancement 29
Quasistatic Properties 29
Fatigue Properties 44
Fatigue Crack Nucleation/Growth 53
Short Crack Behavior: Role of Microstructure 61
Fracture Toughness 63
Probabilistic Fatigue Life Prediction 65
References 69

Chapter 4 Friction Stir Processing: A Potent Property
           Enhancement Tool Viable for Industry 71
Industrial Implementation: Possible Ways of Integrating FSP 71
Impact of FSP on Quality Index of Castings 78
Role of Numerical Tools: FEA of Structures 81
Design Considerations 84
References 84

Chapter 5 Summary and Future Outlook 87

LIST OF FIGURES

Figure 1.1   The nomogram proposed by Drouzy et al.,   3
              illustrating the quality index of a cast Al-7Si-Mg
              type alloy. S_T is the tensile strength of the
              alloy.

Figure 1.2   Shift in the quality index curves as a function of   3
              aging temperature and time.

Figure 2.1   Schematic of FSW process.   6

Figure 2.2   Various microstructural zones in the transverse   6
              cross-section of a friction stir welded material:
              A, parent metal; B, HAZ; C, TMAZ; D, weld
              nugget.

Figure 2.3   Microstructure of the (a) cast A356 plate,   8
              (b) FSP plate.

Figure 2.4   Cross-sectional macrographs showing how changes   9
              in rotation rate and travel speed lead to changes
              in the shape of the process zone. (a) 300 rpm,
              51 mm/min; (b) 300 rpm, 102 mm/min;
              (c) 500 rpm, 51 mm/min; (d) 500 rpm, 102 mm/min;
              (e) 700 rpm, 102 mm/min; (f) 700 rpm, 203 mm/min;
              (g) 900 rpm, 102 mm/min; (h) 900 rpm, 203 mm/min
              (samples were lightly etched).

Figure 2.5   Effect of tool rotation rate on banded structure   10
              formation.

Figure 2.6   Microstructure of the FSP nugget (300 rpm, 51 mm/   10
              min): (a) fine Si particles; (b) coarse Si particles in
              the banded region.

Figure 2.7   Pin designs used for FSP of A356 plates: 11
              (a) standard pin, (b) triflute pin, and
              (c) cone-shaped pin.

Figure 2.8   Change in average Si particle size as a function of   13
              FSP travel speed.

Figure 2.9   Recrystallization related microstructures: **15**
              (a) deformed state, (b) recovered, (c) partially
              recrystallized, (d) fully recrystallized, (e) grain
              growth, and (f) abnormal grain growth.

Figure 2.10   Optical micrographs of the grain structure in an   **16**
              FSP nugget for various processing conditions. All
              four samples were processed at 2236 rpm and
              various tool traverse speeds: (a) 1.00 in/min,
              (b) 2.32 in/min, (c) 5.50 in/min, and
              (d) 8.68 in/min.

Figure 2.11   Grain size distribution in as-FSP condition, cast   **17**
              F357 alloy.

Figure 2.12   Normalized grain size distribution showing   **17**
              log-normal distribution pattern.

Figure 2.13   Configurations for multipass runs: (a) 0% nugget   **20**
              overlap, (b) lateral shift and 50% nugget overlap,
              (c) lateral shift and 50% nugget overlap; for first
              pass to sixth pass, direction of tool travel was
              bottom to top; for seventh pass to twelfth pass,
              direction of tool travel was top to bottom;
              (d) 100% nugget overlap, first pass is the longest
              and sixth pass is the shortest. Curved arrow
              indicates direction of tool rotation. AS, advancing
              side; RS, retreating side.

Figure 2.14   (a) Cross-sectional macrograph of multipass run   **21**
              involving 0% nugget overlap (configuration I);
              (b–d) grain structure of the nugget at various
              locations. AGG has occurred throughout the
              specimen.

Figure 2.15   (a) Cross-sectional macrograph of multipass run   **21**
              involving 50% nugget overlap (configuration II);
              (b–d) grain structure of the nugget at various
              locations. Presence of fine grains is noted from the
              third pass onward.

Figure 2.16   (a) Cross-sectional macrograph of multipass run   **22**
              involving 50% nugget overlap (configuration III);
              (b) and (c) grain structure of the nugget at various
              locations. Presence of fine grains is noted
              throughout the cross-section.

Figure 2.17   Cross-sectional macrographs of multipass run   **22**
               involving 100% nugget overlap (configuration IV).
               (a) First pass, (b) second pass, (c) third pass,
               (d) fourth pass, (e) fifth pass, and (f) sixth pass.
               Area fraction of fine-grained region increases from
               first to sixth pass.

Figure 2.18   FSP run temperature as a function of tool rotation   **23**
               rate. AGG can be prevented when the process
               temperature is below 300 °C, the established
               $T_{c,AGG}$ for the particular study. A356 alloy was
               used.

Figure 2.19   Grain structure after PWHT of single-pass FSP   **24**
               runs, A356 alloy: (a) 300 rpm, no AGG noted;
               (b) 1500 rpm, AGG occurred. Note the difference
               in scale.

Figure 2.20   Macrograph of transverse cross-section of the   **25**
               processed region.

Figure 2.21   Si particle distribution across the process region in a **25**
               multipass FSPed A356 plate. (a) first pass,
               (b) second to third transition zone, and (iii) fifth
               pass (locations A, D, and I in Figure 2.20).

Figure 3.1   Relation between yield strength (YS) and process   **31**
               parameters.

Figure 3.2   Comparison of tensile properties of as-FSPed and   **32**
               as-cast conditions of an F357 alloy: (a) strength and
               (b) ductility.

Figure 3.3   Comparison of tensile properties of FSP + T6   **33**
               and cast + T6 F357 alloy: (a) strength and
               (b) ductility.

Figure 3.4   Tensile stress–strain plots for A206 alloy: effect of **35**
               FSP and heat treatment.

Figure 3.5   Yield strength and ultimate tensile strength (UTS)   **36**
               as a function of ductility; role of various
               heat treatments in increasing yield strength
               is apparent.

Figure 3.6   Microstructure of AZ91 alloy: (a) as-cast (b) FSP   **37**
               nugget.

Figure 3.7   Tensile properties of AZ91 alloy.   **37**

Figure 3.8 β-Mg$_{17}$Al$_{12}$ morphology: (a) and (b) single-pass   **39**
              FSP, (c) pre-ST FSP, and (d) two-pass FSP.
Figure 3.9   Average room temperature tensile properties in an   **41**
              Elektron 21 alloy, as-received (AR) and after FSP
              and subsequent heat treatment schedules, strain rate
              1×10^{-3} s^{-1}.
Figure 3.10   Microstructure of Mg-Gd-Y alloy: (a) Base metal   **42**
              (BM), (b) (SZ) and BM interface, (c) inside SZ,
              (d) SZ at higher magnification.
Figure 3.11   The effect of pore size and shape on fatigue life in   **46**
              an Al-7Si-0.6 Mg casting.
Figure 3.12   The effect of defect size on fatigue life in an   **47**
              A356-T6 casting. The detrimental role of casting
              porosity is clear.
Figure 3.13   S—N plot for A356 alloy.   **48**
Figure 3.14   S—N plot for F357 alloy.   **50**
Figure 3.15   Crack initiation at porosity corner in as-cast F357   **51**
              alloy.
Figure 3.16   Crack initiation at Si/Al-matrix interface, FSP + T6   **52**
              condition, F357 alloy.
Figure 3.17   Defect size comparisons: (a) cast + T6 versus and   **52**
              (b) FSP + T6. ECD, equivalent circular diameter.
Figure 3.18   Crack profile, cast + T6 condition. Arrows show   **53**
              loading direction.
Figure 3.19   Fatigue crack morphology in a cast + T6 F357   **54**
              alloy. Relatively larger Si particles were fractured in
              the crack path.
Figure 3.20   Crack profile in FSP + T6 condition. The crack   **55**
              moved along the particle/matrix interface. Failure
              mode at a particle is either debonding or cleavage,
              based on particle size. Larger arrows indicate
              regions of faster crack growth due to lower particle
              number density. Crack branching is also apparent.
Figure 3.21   Crack initiation and growth in FSP + T6 condition: **57**
              (a) crack initiating in the α-Al matrix after 20,000
              cycles, (b) crack initiation at the particle/matrix
              interface after 20,000 cycles; crack branching along
              grain boundaries is also apparent, (c) dominant

crack is highlighted (black arrow indicates nucleation site whereas red arrow shows crack tip position after 20,000 cycles), (d) crack showing stage I behavior, (e) cracking at later stages in particle/matrix interface away from specimen edge, and (f) crack extension between two observations; crack tip shown in (c) has grown by Δa.

Figure 3.22   Fatigue crack growth rate curve, FSP + T6   **58**
condition, F357 alloy.

Figure 3.23   Change in crack growth path as a function of   **59**
fatigue life: (a) after 40,000 cycles, (b) after 60,000 cycles, (c) after 80,000 cycles, (d) after 100,000 cycles, and (e) after 120,000 cycles. Crack that was growing at 45° to tensile axis is now almost at 90° to the loading direction. Si particles adjacent to the crack have been marked for easier identification of the crack extension event.

Figure 3.24   Crack growth rate curve, cast + T6 versus   **59**
FSP + T6. Dashed line indicates transition to Paris regime.

Figure 3.25   Comparison of SEM fractographs, FSP + T6 (a–c)   **60**
versus cast + T6 (d–f): (b) shows fine striations in FSP + T6 condition, whereas coarse striations are observed in the cast + T6 condition, shown in (e); dimpled rupture morphology is evident in the FSP + T6 sample shown in (c), while the cleavage mode of fracture is visible in the cast + T6 sample, as shown in (f).

Figure 3.26   Oscillatory fatigue crack growth rate behavior,   **61**
F357 alloy.

Figure 3.27   Crack growth rate versus crack length, cast + T6   **63**
versus FSP + T6.

Figure 3.28   Crack growth rates in A356 under various   **64**
processing conditions: (a) $R = 0.1$ and (b) $R = 0.5$.

Figure 3.29   Flow chart of the algorithm used to predict the   **66**
fatigue life.

Figure 3.30   A comparison between the experimental and   **67**
               predicted stress amplitude versus fatigue crack
               initiation life for both cast + T4 and FSP
               conditions. The lower and upper bounds of the
               prediction are shown on the plot.
Figure 3.31   A comparison between the experimental and   **68**
               predicted stress amplitude versus fatigue life for
               both cast + T4 and FSP conditions. The lower and
               upper bounds of the prediction are shown on the
               plot.
Figure 3.32   Stress amplitude versus lower bound of computed   **68**
               fatigue life showing cast + T4 condition (black) and
               FSP conditions from the actual defect distribution
               labeled 1 (blue), the distribution of smaller particles
               labeled 2 (gray), and a condition without particles
               (green). The inset shows the corresponding particle
               distributions marked 1 and 2.
Figure 4.1   Diesel piston top showing fatigue cracks that lead   **72**
               to failure of the bowl rim.
Figure 4.2   FSP applied to an aluminum diesel piston top to   **73**
               improve the fatigue performance of the bowl rim.
Figure 4.3   Typical crankshaft after finish machining showing   **73**
               oil hole locations (a), RBF specimen showing blind
               hole to simulate stress concentration of oil hole
               (b), cross-section of an FSP process specimen
               showing blind hole edges located in the FSP
               microstructure (c).
Figure 4.4   When a hole is drilled in the parent material, the   **74**
               fatigue life drops by a 4 × (a). However if the
               hole is drilled in FSP material, the full life is
               recovered (b).
Figure 4.5   FSP region in cast iron (a). Microstructure showing **77**
               fine-grained TiC reaction products (b). Raster
               pattern of FSP applied to surface and extracted
               rotor after testing (c).
Figure 4.6   A quality index chart for alloy A356. The dashed   **80**
               lines represent the quality index chart as determined
               by Drouzy et al. The solid lines are the flow curves

(identified by the *n*-value), and the iso-*q* lines (identified by *q*-value), calculated with Eqns (4.4) and (4.6), respectively, assuming $K = 430$ MPa.

Figure 4.7   A plot of quality index showing significant in   **81**
              Weibull modulus after friction stir processing of
              A356 alloy.

Figure 4.8   An example of an aluminum wheel showing the   **82**
              solidification pattern and the fatigue stress
              distribution.

Figure 4.9   Log of life contours: (a) wheel rim in cast + T6   **83**
              condition, and (b) wheel rim after FSP + T6 of
              critical regions.

LIST OF TABLES

Table 2.1 Summary of Process Parameters Used for A356 FSP   **8**

Table 2.2 Summary of Tool Pin Design and Process   **11**
Parameters for A356 FSP

Table 2.3 Size and Aspect Ratio of Si Particles in FSP and   **12**
As-Cast A356

Table 2.4 FSP Parameter Details for AGG Study   **20**

Table 2.5 Size and Aspect Ratio of Si Particles in an A356   **26**
Plate Undergoing Five-Pass FSP

Table 3.1 Room Temperature Tensile Properties of As-Cast   **30**
and FSPed A356; Strain Rate = 1×10^{-3} s^{-1}

Table 3.2 Room Temperature Tensile Properties of   **30**
T6-Treated Cast and FSPed A356; Strain
Rate = 1×10^{-3} s^{-1}

Table 3.3 Tensile Properties of AZ80 Alloy   **38**

Table 3.4 Tensile Properties in an HPDC AM60B Alloy after   **39**
FSP

Table 3.5 Tensile Properties of Mg-Gd-Y Alloy Under Various   **42**
Conditions

Table 3.6 Tensile Properties of Cast Mg-Gd-Y Alloy Under   **43**
Various FSP Conditions

Table 3.7 Tensile Properties of FSP Sample   **43**
(800 rpm, 25 mm/min) in the Transverse (TD)
and Longitudinal (LD) Directions

Table 3.8 Process Parameters Used for FSP of As-Cast and   **44**
HIPed Ti6-4 Alloy

Table 3.9 Tensile Properties of As-Cast Ti6-4 Alloy   **44**

Table 3.10 Tensile Properties After FSP of Ti6-4 Alloy   **45**

Table 3.11 Influence of FSP on Particle Size and Porosity   **48**
Volume Fraction in Cast A356

Table 3.12 Comparison of Fatigue Life for Cast and FSPed   **49**
A356 and A319

Table 3.13   Summary of Fracture Toughness Test Data in A356   **64**
             alloy, $R = 0.1$
Table 3.14   Microstructural Features Initiating Cracks, the   **65**
             Number of Cycles for Crack Initiation, and the
             Probability of Occurrence on the Surface
Table 4.1   Impact of Selective FSP Regions on the Fatigue Life **84**
             of Automotive Wheel Rim

Preface to This Volume of the Friction Stir Welding and Processing Book Series

This is the fifth volume in the recently launched short book series on friction stir welding and processing. As highlighted in the preface of the first book, the intention of this book series is to serve engineers and researchers engaged in advanced and innovative manufacturing techniques. Friction stir welding was invented more than 20 years back as a solid state joining technique. In this period, friction stir welding has found a wide range of applications in joining of aluminum alloys. Although the fundamentals have not kept pace in all aspects, there is a tremendous wealth of information in the large volume of papers published in journals and proceedings. Recent publications of several books and review articles have furthered the dissemination of information.

This book is focused on friction stir casting modification, an area that has unique potential to embed wrought microstructure in cast components. It is very well established that the wrought microstructure results in best properties whereas casting is the most cost-effective way of making complex shaped components. The basic research papers continue to establish the performance advantages of this approach. The early implementation of technology is still lacking and hopefully this book will provide confidence to designers and engineers to consider friction stir casting modification for a wider range of applications. It will also serve as a resource for researchers dealing with various aspects of friction stir casting modification concept. As stated in the previous volumes, this short book series on friction stir welding and processing will include books that advance both the science and technology.

Rajiv S. Mishra
University of North Texas
September 6, 2015

CHAPTER 1

Introduction

Casting is one of the oldest and most direct shaping operations known to mankind. In casting, liquid metal is poured into a hollow cavity (the mold), which is dimensionally very close to the finished form. Once solidified, the part is removed from the mold, and may need a few post-processing steps (machining, heat treatment, etc.) before it is ready for its intended use. Metal casting is highly flexible in terms of configuration design, and if a pattern can be made for a part, the part can be cast. The ability to produce simple and/or complex, intricate shapes economically makes casting a highly attractive manufacturing process. Casting is commonly used for infrastructure and structural components, water distribution systems, automotive components, prosthetics, jewelry, and gas turbine engine hardware [1]. A 2005 survey by *Modern Casting* reported worldwide annual casting production to be just less than 80 million metric tons. The latest survey shows a total annual worldwide casting production of more than 100 million metric tons at the end of 2012, representing ~4% growth per annum in this sector over the last 7 years [2].

However, the directness of the casting process as a manufacturing operation also makes it quite complex to manage. The major reason is linked to the solidification process that takes place inside a mold. A good casting design needs to consider (i) fluid life/fluidity, (ii) solidification shrinkage, (iii) solid shrinkage, (iv) slag/dross formation, (v) pouring temperature, (vi) fluid flow, and other characteristics to produce a quality casting [1]. Based on the alloy being cast, the conditions can be very different for each of the abovementioned criteria. For example, A356 (Al-7Si-Mg type alloy) has low pouring temperature, excellent fluid life, and mostly eutectic-type solidification behavior, which makes this alloy highly amenable to casting. On the other hand, carbon steel has high pouring temperature, low fluidity, and large directional-type solidification shrinkage, making it much more challenging to cast. These variables make chances of having internal defects in a casting fairly high, especially

Friction Stir Casting Modification for Enhanced Structural Efficiency. DOI: http://dx.doi.org/10.1016/B978-0-12-803359-3.00001-7

if the design criteria were not addressed rigorously. The chances of encountering a defect could be even higher based on the nature of the solidification of the alloy being cast. Solidification shrinkage with gas entrapped in the liquid melt can lead to porosity. Turbulent fluid flow and slag/dross formation can result in trapped oxide bifilms—inclusions inside a casting.

Structural defects, i.e., pores and oxide bifilms, significantly degrade mechanical properties. The presence of such defects in a casting leads to premature fracture in tension and fatigue, and results in low ductility, tensile strength, and fatigue life [3,4]. Moreover, the presence of casting defects that cause significant variability in mechanical properties eliminates the possibility of using castings in critical applications in the automotive and aerospace sector. Minimization and possible elimination of casting defects is, therefore, desired for wider use of castings as structural components. Concern about variability in mechanical properties and structural integrity of a casting led to the development of various quality indices for aluminum alloy castings [5]. The use of a quality index provides the foundry engineer with a metric for structural quality as well as the effects of any casting process improvement on the mechanical properties of the casting. Drouzy, Jacob, and Richard (DJR) were the first to introduce an empirical equation that describes the relationship between yield strength (σ_Y), tensile strength (σ_{TS}), and elongation to failure (e_F) in Al-7Si-Mg alloy castings, as shown below [6]:

$$\sigma_Y = \sigma_{TS} - a \log_{10} e_F - b \tag{1.1}$$

where a is 60 MPa and b is 13 MPa. Equation (1.1) is valid for $e_F \geq 1\%$. A linear relationship between σ_{TS} and $\log_{10}(e_F)$, as observed by DJR, led to the first proposed quality index, Q_{DJR}. For under-aged and peak-aged alloys, Q_{DJR} is expressed as follows:

$$Q_{DJR} = \sigma_{TS} + 150 \log_{10} e_F \tag{1.2}$$

Drouzy et al. provided the nomogram, which is presented as Figure 1.1, on which the three tensile properties can be plotted [6]. The nomogram has contour lines of constant σ_Y and of constant Q_{DJR}. The effects of several process variables, i.e., minor changes in alloy chemistry, heat treatment, solidification rate, etc., can be reflected on a quality index plot very succinctly. One such example is shown in Figure 1.2, where influences of aging parameters are shown for a cast A354 (Al-9Si-Cu-Mg) type alloy [7]. The shift in the quality index curves as the aging temperature is increased is apparent. Other quality indices have also been used. For a more complete review, readers can refer to the resources listed in the references section.

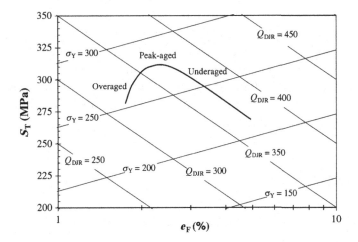

Figure 1.1 The nomogram proposed by Drouzy et al., illustrating the quality index of a cast Al-7Si-Mg type alloy. S_T is the tensile strength of the alloy.

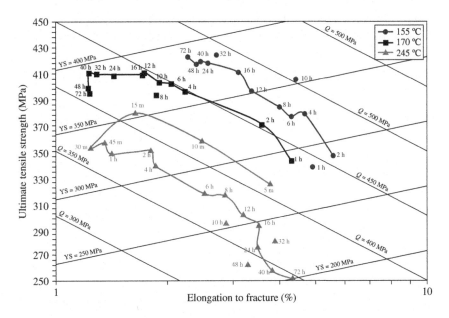

Figure 1.2 Shift in the quality index curves as a function of aging temperature and time [7].

It appears from Eqn (2) that the quality of a given Al casting would improve with concurrent improvement in tensile strength and tensile ductility. Normally, this is done by optimizing the casting process, i.e., change in alloy chemistry, modification of mold design, post-process heat treatment, etc. However, in this short book series, we are going to discuss how friction stir processing (FSP), a novel surface processing technique, can enhance the quality of a given casting. Modification of castings by FSP for different types of casting alloys has been attempted by several researchers, and the results are very encouraging. A concurrent increase in tensile strength and tensile ductility is apparent after FSP, together with various other microstructural benefits. These will be discussed over the next chapters of this book.

REFERENCES

[1] Casting. In: ASM handbook, vol. 15; 2008. p. 1–13.

[2] 47th Census of World casting production, Modern casting; December 2013.

[3] Tiryakioglu M, Campbell J, Staley JT. Scripta Mater 2003;49:873–8.

[4] Nyahumwa C, Green NR, Campbell J. Metall Mater Trans A 2001;32A:349–58.

[5] Tiryakioglu M, Campbell J, Alexopoulos ND. Metall Mater Trans B 2009;40B:802–11.

[6] Drouzy M, Jacob S, Richard M. AFS Intl Cast Metals J 1980;5:43–50.

[7] Ammar HR, Samuel AM, Samuel FH, Simielli E, Sigworth GK, Lin JC. Metall Mater Trans A 2012;43A:61–73.

CHAPTER 2

Friction Stir Processing: An Introduction

Friction stir processing (FSP) is an adaptation of the friction stir welding (FSW) technique, originally invented at The Welding Institute, United Kingdom in 1991 [1]. During FSW, a rotating nonconsumable tool with a specially designed pin and shoulder is first inserted into the joint line. This is the plunge step. Once the desired plunge depth is reached, the tool travels along the joint line to complete the weld. A schematic of the process is shown in Figure 2.1.

The spinning action of the FSW tool results in significant frictional heating between the metal workpiece and the tool shoulder. Additionally, further adiabatic heating results from severe plastic deformation of the metal workpiece. Tremendous localized heating at the tool/workpiece interface leads to softening of the material around the pin. However, material just outside the tool pin/shoulder remains at relatively much lower temperature, thereby constraining the plasticized material. As the FSW tool moves forward, rotation coupled with translation of the tool causes this soft material to flow around the pin under the shoulder from the leading edge of the tool to the trailing edge of the rotating tool, thus filling the hole in the wake of the tool. Once the soft metal flows and comes behind the pin, the tool shoulder forges it down, completing the weld. The most notable feature of the FSW is the generation of a defect-free solid-state joint, i.e., without any melting, when performed normally. Thus, it is clear that FSW utilizes the high plasticity of material at elevated temperatures for the metal joining process. Significant changes occur in the weld zone around the pin as a result of the tool movement. As has been mentioned in one of the recent reviews on FSW [2], the weld zone can be divided into following four regions, shown as a schematic in Figure 2.2:

- *Unaffected material or parent metal:* This corresponds to a region that is far enough away from the weld zone that no microstructural or mechanical changes occur.

Friction Stir Casting Modification for Enhanced Structural Efficiency. DOI: http://dx.doi.org/10.1016/B978-0-12-803359-3.00002-9

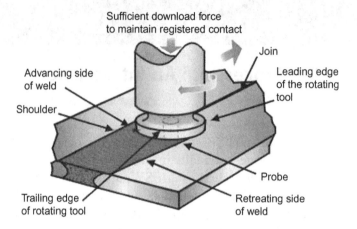

Figure 2.1 Schematic of FSW process.

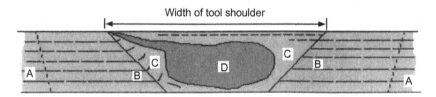

Figure 2.2 Various microstructural zones in the transverse cross-section of a friction stir welded material: A, parent metal; B, HAZ; C, TMAZ; D, weld nugget [2].

- *Heat affected zone (HAZ):* This region is closer to the weld zone. Due to the proximity to the heat source, microstructural changes cause changes in mechanical properties. There is no plastic deformation in this region.
- *Thermomechanically affected zone (TMAZ):* This region corresponds to locations that have undergone deformation and also have been affected by heat. There is no recrystallization in this region. This is located very near to the pin.
- *Weld nugget:* This is the region where material undergoes complete recrystallization, and as a result fine equiaxed grain structure evolves. This corresponds to areas that were occupied by the pin previously. This zone is also sometimes referred to as the stir zone (SZ).

During FSP, exactly the same procedure is carried out except that FSP is carried out on a bulk component. Instead of creating a joint line between two separate pieces of metal, FSP is carried out on a single piece of metal at specific regions of interest, targeting localized modification of microstructure. As FSP results in flow of plasticized material around the tool, a wrought microstructure is created along the processed zone. Mishra et al. [3,4] were the first to attempt the FSP technique in a commercial 7075 Al alloy. Their work showed how formation of a uniform fine-grain microstructure in the FSP zone results in significantly enhanced high-strain-rate superplasticity. As compiled in a recent review by Ma [5], FSP offers several benefits. First, FSP is a short-route, solid-state processing technique. In a single processing step, it can produce microstructural refinement, densification, and homogeneity. Second, the microstructure and mechanical properties of the processed zone are functions of the thermomechanical processing history during FSP. Therefore, they can be controlled by optimizing the tool design, FSP parameters, and active cooling/heating. Third, the depth of the processed zone can be adjusted by changing the length of the tool pin. This is a tremendous benefit, because with other surface processing techniques it is very difficult to adjust the depth of the desired process zone. FSP is also less energy intensive than other techniques, making it a "green" technology; this is another significant advantage.

As the rotation of the FSP tool pin results in intense breaking and mixing of material inside the process zone (nugget), FSP has the ability to create a refined, uniform microstructure in a typically heterogeneous material, such as cast metallic alloy plates. The possible development of FSP as a generic microstructural modification tool specifically aimed at cast materials is described in detail in the following sections of this book.

MICROSTRUCTURE: EFFECT OF PROCESS PARAMETERS

FSP leads to significant breakdown of cast microstructure. However, the resulting microstructure inside the FSP nugget is intimately linked with the process parameters used, namely tool rotation rate and tool travel speed. Design of the FSP tool plays an equally important role in determining the microstructure. Ma et al. [6] carried out a comprehensive study to understand the effects of FSP process parameters on evolution of microstructure inside the nugget. A commercial cast A356 Al billet (Al-7Si-0.3Mg, wt%) was used in this study. Single-pass FSP was performed on 6.35-mm thick A356 plates machined from as-received cast billets. Table 2.1 summarizes the details of process parameters

Table 2.1 Summary of Process Parameters Used for A356 FSP [6]			
Tool Rotation Speed (rpm)	Tool Travel Speed (mm/min)		
	51	102	203
300	x	x	
500	x	x	
700		x	x
900		x	x

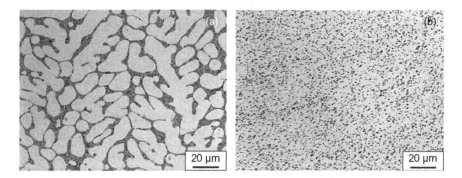

Figure 2.3 Microstructure of the (a) cast A356 plate, (b) FSP plate [6].

used in this work. After FSP, the processed plates were sectioned for microstructural characterization using various microscopy techniques.

Ma et al. [6] showed how FSP dramatically modified the cast microstructure. Microstructure of the as-received cast billet is shown in Figure 2.3a, which comprises primary α-Al dendrites and irregular Si particles in interdendritic regions. FSP led to elimination of the dendritic structure and refinement and uniform distribution of Si particles inside the nugget, as shown in Figure 2.3b. The role of process parameters on microstructural evolution is captured in Figure 2.4. The series of macrographs shows how shape of the nugget changes as a function of FSP process parameters. Lower tool rotation rates (300–500 rpm) result in basin-shaped nuggets with wide top regions (Figure 2.4a–d). Increase in tool rotation rate changes the nugget shape to elliptical (Figure 2.4e–h), and the formation of an "onion ring" structure has begun. A totally elliptical nugget with onion ring structure is formed at 900 rpm and 203 mm/min of travel speed. Onion ring structures are found on the transverse cross-section of most FSW joints. As the name suggests, the structure consists of a series of

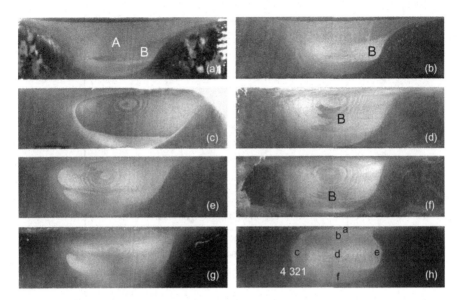

Figure 2.4 Cross-sectional macrographs showing how changes in rotation rate and travel speed lead to changes in the shape of the process zone. (a) 300 rpm, 51 mm/min; (b) 300 rpm, 102 mm/min; (c) 500 rpm, 51 mm/min; (d) 500 rpm, 102 mm/min; (e) 700 rpm, 102 mm/min; (f) 700 rpm, 203 mm/min; (g) 900 rpm, 102 mm/min; (h) 900 rpm, 203 mm/min (samples were lightly etched) [6].

concentric circles with the center located near the middle of the nugget, and the spacing between the circles wider at the center and narrower toward the weld edges. Krishnan [7] studied the formation of onion rings in detail and concluded that the structure is a geometric effect that is linked with simultaneous rotation and forward movement of the FSW tool and associated material flow around the tool pin. The spacing is related to weld pitch, i.e., tool advance per revolution.

Additionally, Ma et al. [6] noted formation of a banded structure (region B, Figure 2.4) on transverse section macrographs of the process zone for certain parameter combinations. The banded structure appeared to be forming toward the advancing side, when lower tool rotation rates were used. At higher tool rotation rates, banded structure was not noticed. To provide further insight into banded structure formation, Ma et al. [6] plotted tool rotation rate versus tool travel speed (Figure 2.5). This plot shows that increasing either the tool rotation rate or the rotation rate/travel speed ratio eliminates banded structure. The effect of FSP on Si particle size and morphology was carried out by optical microscopy at a higher magnification. The presence of fine and equiaxed Si particles inside most of the nugget (Figure 2.6a) is

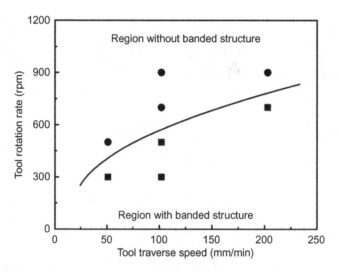

Figure 2.5 Effect of tool rotation rate on banded structure formation [6].

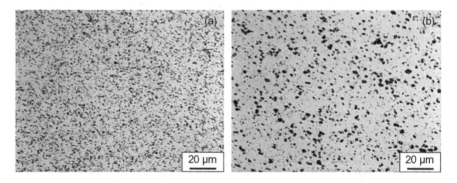

Figure 2.6 Microstructure of the FSP nugget (300 rpm, 51 mm/min): (a) fine Si particles; (b) coarse Si particles in the banded region [6].

apparent. However, examination of the banded region reveals a low density of coarse particles (Figure 2.6b). The interface between the banded region and the rest of the nugget was found to be fully bonded. The study by Ma et al. [6] shows the importance of using a relatively higher tool rotation rate in order to create a nugget with uniform, homogeneously distributed fine Si particles in a cast A356 type of alloy.

In a separate study, Ma et al. [8] applied FSP to sand-cast A356 alloy plates in which Si particles were not chemically modified by Na/Sr treatment. As a result, Si particles were long and had a needle-like

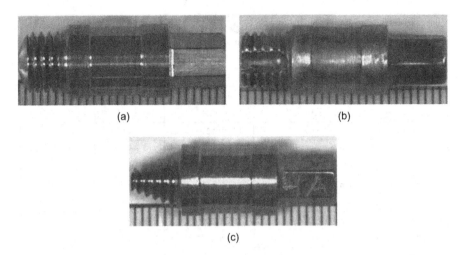

(a)   (b)

(c)

Figure 2.7 Pin designs used for FSP of A356 plates: (a) standard pin, (b) triflute pin, and (c) cone-shaped pin [8].

Table 2.2 Summary of Tool Pin Design and Process Parameters for A356 FSP [8]			
Processing Parameter	**Tool Geometry**		
	Standard Pin (S-Pin)	**Triflute Pin (T-Pin)**	**Cone-Shaped Pin (C-Pin)**
300 rpm–51 mm/min	x	x	x
700 rpm–203 mm/min	x	x	x
900 rpm–203 mm/min	x	x	–
1100 rpm–203 mm/min	x	x	–

morphology. This study examined the effects of both FSP process parameters and FSP tool design on nugget microstructural evolution. For the FSP tool, three different pin designs were adopted: (i) standard pin, (ii) triflute pin, and (iii) conical pin (Figure 2.7). Table 2.2 shows the FSP parameters and tool designs that were used in this study. FSP resulted in breakup of dendritic structure and significant refinement in the size and shape of Si particles. Details about the Si particle size and shape are summarized in Table 2.3. In general, FSP reduced the average size of Si particles that were present in the as-cast condition by more than a factor of five. Aspect ratio is lower by a factor of three after FSP. Casting porosity is also dramatically reduced after FSP. On closer examination of the particle size and shape data, the effect of tool rotation rate is apparent. Increasing tool rotation rate from 300 to 1100 rpm led to reduction in both the size and the aspect ratio of Si particles as well as porosity content. Microstructural study revealed

Table 2.3 Size and Aspect Ratio of Si Particles in FSP and As-Cast A356 [8]			
Material	Particle Size (μm)	Aspect Ratio	Porosity Volume Fraction (pct)
As-cast	16.75 ± 9.21	5.92 ± 4.34	0.95
FSP, 300 rpm−51 mm/min (S-pin)	2.84 ± 2.37	2.41 ± 1.33	0.11
FSP, 700 rpm−203 mm/min (S-pin)	2.62 ± 2.31	1.93 ± 0.86	0.050
FSP, 900 rpm−203 mm/min (S-pin)	2.55 ± 2.21	2.00 ± 1.01	0.027
FSP, 1100 rpm−203 mm/min (S-pin)	2.51 ± 2.00	2.04 ± 0.91	0.042
FSP, 300 rpm−51 mm/min (T-pin)	2.70 ± 2.26	2.30 ± 1.15	0.087
FSP, 300 rpm−51 mm/min (T-pin)-T6	3.69 ± 2.55	1.93 ± 0.90	−
FSP, 700 rpm−203 mm/min (T-pin)	2.50 ± 2.02	1.94 ± 0.88	0.024
FSP, 700 rpm−203 mm/min (T-pin)-T6	3.05 ± 2.04	1.59 ± 0.55	−
FSP, 900 rpm−203 mm/min (T-pin)	2.50 ± 2.04	1.99 ± 0.94	0.032
FSP, 900 rpm−203 mm/min (T-pin)-two-pass	2.43 ± 2.02	1.86 ± 0.78	0.020
FSP, 1100 rpm−203 mm/min (T-pin)	2.44 ± 2.00	1.86 ± 0.81	0.025
FSP, 300 rpm−51 mm/min (C-pin)	2.90 ± 2.46	2.50 ± 1.35	0.094
FSP, 700 rpm−203 mm/min (C-pin)	2.86 ± 2.32	2.09 ± 0.90	0.032
S, standard; T, triflute; C, cone shaped.			

the presence of a larger quantity of fine Si particles when processed at a higher tool rotation rate than at a slower tool rotation rate. Regarding pin geometry, the triflute pin design appeared to be more effective than the standard pin or the cone-shaped pin in refining the large needle-shaped Si particles.

Similar study by Santella et al. [9] on A319 (Al-6Si-3.5Cu, wt%) and A356 alloys also established the beneficial action of FSP on microstructure: (i) closure of casting porosity, (ii) refinement of large second-phase particles, (iii) uniform distribution of second-phase particles, and (iv) removal of the dendritic microstructural pattern formed during solidification. Lee et al. [10] showed the apparent effect of FSP travel speed on second-phase particle size. FSP was performed on A356 plates at a fixed tool rotation rate (1600 rpm) at different travel speeds (87−342 mm/min). The average size of Si particles increased slightly from 6.46 to 7.89 μm as the travel speed increased from 87 to 267 mm/min (Figure 2.8). Lee et al. [10] concluded that, at a given tool rotation rate, use of a slower travel speed results in more refined and uniformly distributed Si particles inside the SZ. Jana et al. [11] in their study on a cast F357 alloy (Al-7Si-0.6Mg, wt%) similarly concluded that a higher volume fraction of finer second-phase particles was likely

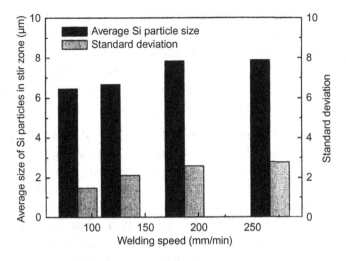

Figure 2.8 Change in average Si particle size as a function of FSP travel speed [10].

to be produced when high tool rotation rate and low-to-intermediate travel speeds were used. The effect of using higher tool rotation rate in obtaining more refined second-phase particles in cast Al-Si alloys has been confirmed by other researches as well [12,13].

In summary, FSP tool rotation rate plays a major role in refining second-phase particles that may be present in a cast metallic alloy system. Use of a higher rotational speed will have greater particle refining effect and will also lead to more homogeneous particle distribution inside the process zone. Tool travel speed plays a somewhat secondary role. Increased particle refinement is observed at low-to-intermediate tool travel speeds. In addition, FSP results in closure of casting porosity and removal of dendritic microstructural patterns.

RECRYSTALLIZATION MECHANISMS

In "Microstructure: effect of process parameters" section, the role of FSP process parameters in refining the size and shape of second-phase particles in cast metallic alloys was discussed in detail. However, as mentioned earlier, apart from refining second-phase particles, FSP also converts the dendritic matrix microstructure into a wrought microstructure consisting of equiaxed grains. Recrystallization mechanisms play a critical role in determining the final microstructure of the SZ. FSP leads to severe plastic deformation (SPD) of the dendritic matrix microstructure, and, as a result, introduction of various microstructural defects (dislocations, interfaces, etc.) takes place. As FSP involves simultaneous friction/deformation-induced heating (a thermomechanical process, TMP), restoration of the deformed microstructure ensues through thermally activated processes such as solid-state diffusion and dislocation substructure formation.

According to Humphreys and Hatherly [14], two mechanisms, namely (i) recovery and (ii) recrystallization, play a vital role in restoring the deformed microstructure. Recovery involves annihilation and rearrangement of dislocations, which takes places relatively homogeneously throughout the entire deformed matrix and does not usually affect the boundaries between deformed grains. The dislocation structure is not completely removed during the recovery process. Recrystallization, on the other hand, involves nucleation of new dislocation-free grains within the deformed or recovered structure. These new grains then grow thorough large-scale grain boundary migration, resulting in a microstructure with low dislocation density. Either of these two processes can occur during deformation, and are termed dynamic recovery or dynamic recrystallization. When exposed to elevated temperature, as occurs during FSP, a recrystallized microstructure can further undergo normal grain growth, where larger grains grow at the expense of smaller grains. In a few instances, normal grain growth might be replaced by a process called abnormal grain growth (AGG) or secondary recrystallization, where selected few grains grow to a very large size. Grain growth would eventually lead to coarsening of the microstructure. The process of recovery, recrystallization, and grain growth is schematically represented in Figure 2.9.

Jana et al. [11], in their study on FSP of F357 alloy, attempted to reveal the matrix grain structure by chemical etching. Examples of grain structure in as-FSPed condition are shown in Figure 2.10, which were obtained from samples processed at 2236 rpm and various traverse speeds. FSP led to significant grain refinement inside the nugget due to intense plastic deformation. The grains were mostly equiaxed, although in a few instances some grains aligned parallel to either the advancing or retreating side. Higher tool traverse speeds appear to produce slightly higher fractions of finer grains, as is evident from a comparison of Figure 2.10d with the other three images. All four processing conditions produced a similar mean grain size intercept value

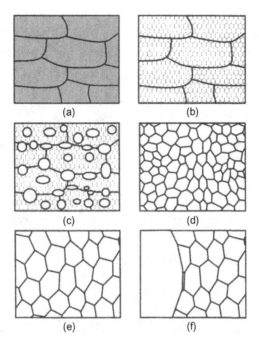

Figure 2.9 Recrystallization related microstructures: (a) deformed state, (b) recovered, (c) partially recrystallized, (d) fully recrystallized, (e) grain growth, and (f) abnormal grain growth [14].

of ~ 6 μm, suggesting that the range of process parameters used has secondary effects on average grain size.

In the same study [11], direct measurements of individual grains were carried out, followed by averaging, rather than the usual "mean linear intercept" technique for estimating grain size. As a result, additional information about grain size distribution was acquired. Figure 2.11 shows grain size distribution for the 2236 rpm processing in the as-FSPed condition. The plot shows the presence of a slightly higher fraction of smaller grains and a narrower range of grain size at a tool traverse speed of 8.68 in/min—a result of restricted grain growth due to the faster cooling rate associated with higher travel speed. In Figure 2.12, the same grain size distribution is shown as a function of normalized grain size (R/R_{mean}). This plot indicates that grain size distribution for the present study followed either a log-normal or Rayleigh function. According to Humphreys and Hatherly [14], grain size distribution is closest to a Rayleigh function, although Su et al. [15] concluded that grain size distribution in the SZ follows a log-normal function.

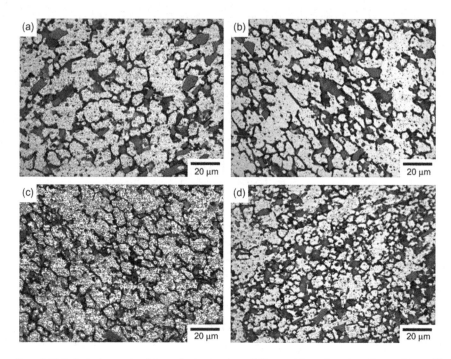

Figure 2.10 Optical micrographs of the grain structure in an FSP nugget for various processing conditions. All four samples were processed at 2236 rpm and various tool traverse speeds: (a) 1.00 in/min, (b) 2.32 in/min, (c) 5.50 in/min, and (d) 8.68 in/min [11].

Alternative views exist about the operative mechanisms that are responsible for generation of fine grains inside an FSPed nugget. They can be broadly classified into two groups: continuous dynamic recrystallization (CDRX)-based mechanisms and discontinuous dynamic recrystallization (DDRX)-based mechanisms. As noted by Humphreys and Hatherly [14], continuous recrystallization is a special case of extended recovery. The subgrains that form during recovery coarsen and finally become comparable to those of a normal grain structure through the process of extended recovery. There is no evidence, however, that the resulting boundaries are predominantly high angle, or that new high-angle grain boundaries (HAGBs) are created during the process. According to Humphreys and Hatherly [14], the difference between extended recovery and continuous recrystallization lies in the relative proportions of low and high-angle grain boundaries in the microstructure. Therefore, if a process that starts with extended recovery finally produces a microstructure composed essentially of HAGBs, this process can be called continuous recrystallization. The prefix

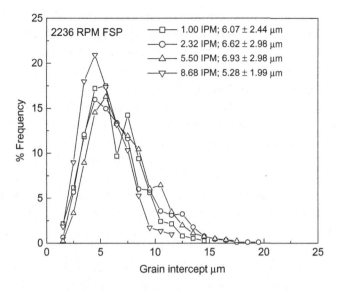

Figure 2.11 Grain size distribution in as-FSP condition, cast F357 alloy [11].

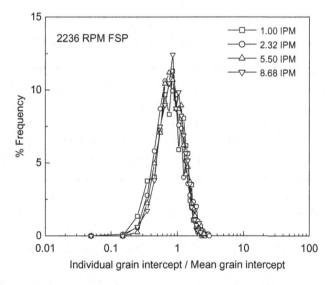

Figure 2.12 Normalized grain size distribution showing log-normal distribution pattern [11].

"dynamic" is added when this process occurs during deformation itself. Discontinuous recrystallization, on the other hand, is characterized by the nucleation of strain-free grains and the growth of these grains through large-scale migration of HAGBs.

Support for CDRX-based mechanisms was first offered by Jata and Semiatin [16], who pointed out that grain structure evolution during FSP occurs through a continuous process in which existing subgrains rotate during deformation to a high angle of misorientation, finally becoming separate recrystallized grains. Su et al. [17] and Fonda et al. [18] came to similar conclusions, although both indicated that the evolution of a new set of much smaller subgrains was responsible for the final recrystallized microstructure, rather than the existing subgrains. Formation of very small subgrains with low-angle boundaries takes place intragranularly through dynamic recovery during the initial phase of FSP as dislocation density and process temperature increase. Subsequent stages of deformation during FSP lead to further introduction of dislocations, which are absorbed by the subgrain boundaries to maintain strain compatibility. There is rotation of subgrains as well. Repeated absorption and rotation of subgrains leads to formation of recrystallized equiaxed grains with high-angle boundaries. Prangnell and Heason [19] showed that geometric dynamic recrystallization (GDR), one of the CDRX mechanisms, was responsible for fine-grain evolution in the SZ. Further support for CDRX-based mechanisms is found in the fact that high stacking fault energy in aluminum alloys makes dynamic recovery rapid and extensive at high temperature [14].

Evidence supporting DDRX-based mechanisms has also been reported recently [15,20–22]. Rhodes et al. [20] observed the presence of extremely fine grains (50–100 nm) at the bottom of the plunge hole that were separated by HAGBs. In a series of experiments, Su et al. [15,21,22] also demonstrated the presence of extremely fine grains with HAGBs and dislocation-free grain interiors around the pin. According to Su et al. [22], a complex state of stress as well as strain components with large strain gradients are produced in the processed material during FSP. Additionally, a large dislocation density is introduced to accommodate strain incompatibility across grains. The combination of the complex state of stress, inhomogeneous strain pattern, and the presence of a very high density of geometrically necessary dislocations (GNDs) results in a high rate of nucleation. Such high rate of nucleation plays a critical role in recrystallization of the FSP nugget through DDRX-based routes.

Interestingly, the fine-grained FSP nugget microstructure shows a rather curious grain growth phenomenon when exposed to elevated temperature after FSP. The fine grain microstructure is typically replaced by a very coarse grain structure, with each grain becoming as large as ~1 mm. It has been found that the evolution of such a coarse grain microstructure, otherwise known as AGG or secondary recrystallization, after exposing the FSP nugget to elevated temperature is inherently linked with the FSP parameters.

Jana et al. [23] used various combinations of FSP parameters to study AGG behavior in a cast F357 alloy. FSP of the cast plates was carried out in single-pass as well as multiple-pass configurations. "Multiple pass" implies FSP of already processed material. It is defined in terms of nugget overlap. A 100% nugget overlap means the processed nugget (obtained after the first run) is being fully reprocessed during subsequent FSP runs. A partial nugget overlap, on the other hand, indicates shifting of the FSP tool to either the advancing or the retreating side after the first FSP run is carried out. For example, a 50% nugget overlap indicates that the FSP tool was shifted so that 50% of the previously formed nugget was reprocessed during subsequent runs. Details of the FSP parameters used in the study by Jana et al. [23] are listed in Table 2.4. Figure 2.13 is a schematic representation of the various multiple-pass configurations employed in this study. Grain growth characteristics of differently processed nuggets were subsequently determined by annealing the processed coupons at 540 °C for 8 h.

Signs of AGG were noted for all the single-pass runs in this study. However, a change in AGG behavior was noted for the various multiple-pass runs. Figure 2.14a shows a cross-sectional macrograph of the 0% nugget overlap condition after high-temperature exposure. Occurrence of AGG is clearly visible in Figure 2.14b−d. For the 50% nugget overlap situation (configuration II), the FSP tool was laterally shifted by 2.5 mm toward the retreating side. Figure 2.15a shows a macrograph of the nugget cross-section from this configuration after high-temperature exposure. Presence of very fine grains of ~10 μm (Figure 2.15b−d) was confirmed at several locations in the processed nugget. It was also observed that the area fraction of the fine-grained region increased from the fourth to the sixth pass. Multiple passes in the third configuration (Figure 2.13c) involved doubling the process runs. For the first six passes, the tool was moved from the bottom of the image to the top, with 2.5 mm of tool lateral shift between runs.

Table 2.4 FSP Parameter Details for AGG Study [23]					
FSP Run Configuration		**Tool Rotation Rate, Tool Travel Speed (rpm, mm/s)**			
Single pass		**2236, 0.42**	**2236, 0.98**	**2236, 2.33**	**2236, 3.67**
Multiple pass (% nugget overlap)	Configuration I (0%)	2236, 2.33			
	Configuration II (50%)	2236, 2.33		1500, 2.33	
	Configuration III (50%)	2236, 2.33		1500, 2.33	
	Configuration IV (100%)	2236, 2.33			

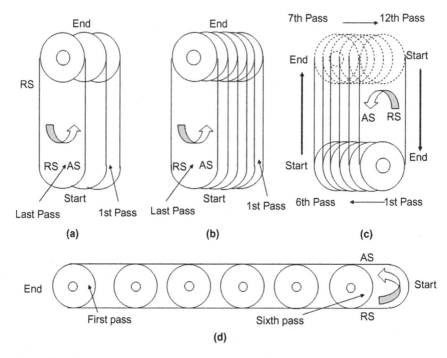

Figure 2.13 Configurations for multipass runs: (a) 0% nugget overlap, (b) lateral shift and 50% nugget overlap, (c) lateral shift and 50% nugget overlap; for first pass to sixth pass, direction of tool travel was bottom to top; for seventh pass to twelfth pass, direction of tool travel was top to bottom; (d) 100% nugget overlap, first pass is the longest and sixth pass is the shortest. Curved arrow indicates direction of tool rotation. AS, advancing side; RS, retreating side [23].

For passes 7–12, the tool was moved from the top of the image to the bottom. As Figure 2.16a shows, the nugget proved to be the least susceptible to AGG when the cast material was processed several times by subsequent FSP runs. The nugget is mostly comprised of fine grain microstructure (Figure 2.16b and c). Similar behavior in AGG pattern

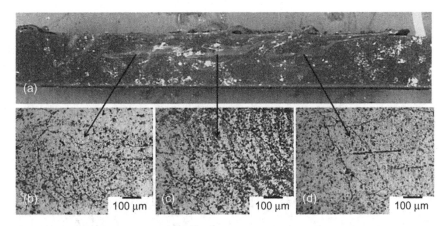

Figure 2.14 (a) Cross-sectional macrograph of multipass run involving 0% nugget overlap (configuration I); (b–d) grain structure of the nugget at various locations. AGG has occurred throughout the specimen [23].

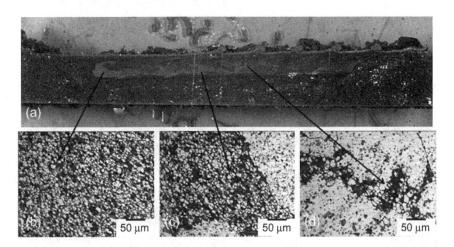

Figure 2.15 (a) Cross-sectional macrograph of multipass run involving 50% nugget overlap (configuration II); (b–d) grain structure of the nugget at various locations. Presence of fine grains is noted from the third pass onward [23].

was observed for configuration IV (Figure 2.13d), where 100% nugget overlap was employed. The volume fraction of fine grains rises with increase in the number of FSP runs (Figure 2.17a–f). Use of a slightly lower tool rotation rate (1500 rpm) in configurations II and III, however, showed occurrence of AGG after high temperature exposure.

The study by Jana et al. [23] clearly establishes the role of multiple passes in controlling nugget grain growth when exposed to elevated temperature. During FSP, the tool carries material from the advancing

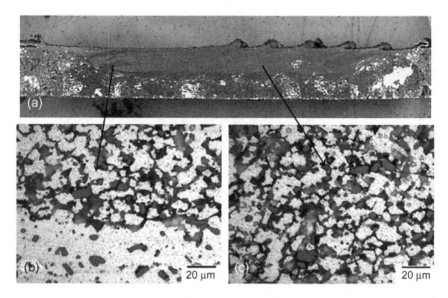

Figure 2.16 (a) Cross-sectional macrograph of multipass run involving 50% nugget overlap (configuration III); (b) and (c) grain structure of the nugget at various locations. Presence of fine grains is noted throughout the cross-section [23].

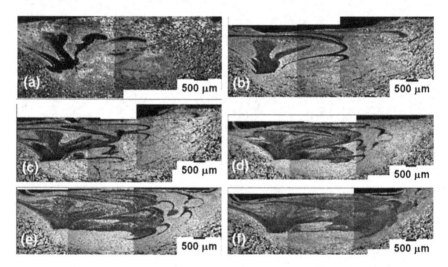

Figure 2.17 Cross-sectional macrographs of multipass run involving 100% nugget overlap (configuration IV). (a) First pass, (b) second pass, (c) third pass, (d) fourth pass, (e) fifth pass, and (f) sixth pass. Area fraction of fine-grained region increases from first to sixth pass [23].

side toward the retreating side, and as a result the material becomes strained. However, the strain distribution is inhomogeneous and a strain gradient is present from the advancing to the retreating side. With repeated processing, the inhomogeneity decreases and overall

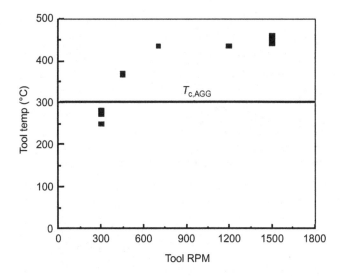

Figure 2.18 FSP run temperature as a function of tool rotation rate. AGG can be prevented when the process temperature is below 300 °C, the established $T_{c,AGG}$ for the particular study. A356 alloy was used [24].

strain increases. After multiple-pass runs, both homogeneity and increased strain inside the nugget lead to elimination of AGG.

A recent FSP study by Jana and Grant [24] on cast A356 alloy plates reveals that FSP process temperature plays an equally important role in determining the stability of the nugget microstructure when the nugget is exposed to elevated temperature. This study involved single-pass runs. Four tool rotation rates, namely 300, 700, 1200, and 1500 rpm, at a fixed tool travel speed, 4 in/min, were used. Temperature during FSP runs was recorded by placing a thermocouple behind the tool shoulder. Process temperature as a function of tool rotation rate is shown in Figure 2.18. Monotonic increase in process temperature with increase in tool rotation rate is evident, which is a characteristic of any FSW/P run. The process temperature is noted to be above 300 °C except with the 300-rpm condition. To study grain growth behavior, a postweld heat treatment (PWHT) at 535 °C for 2.5 h was carried out. All FSP processing conditions led to AGG after PWHT except those carried out at a tool rotation rate of 300 rpm. Microstructure from a part of the nugget after PWHT is shown in Figure 2.19a and b. The presence of fine grain structure for the 300-rpm condition is apparent in Figure 2.19a, while the 1500-rpm condition clearly led to AGG (Figure 2.19b). It is likely that there is a

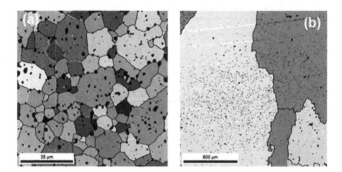

Figure 2.19 Grain structure after PWHT of single-pass FSP runs, A356 alloy: (a) 300 rpm, no AGG noted; (b) 1500 rpm, AGG occurred. Note the difference in scale [24].

critical process temperature, $T_{c,AGG}$ (300 °C for the present case) associated with the occurrence of AGG. To prevent AGG, the key is to form a volumetric defect-free nugget while keeping the FSP temperature below this critical temperature.

In summary, AGG in FSP microstructure after PWHT is a common feature that leads to very coarse grains inside the nugget. However, occurrence of AGG can be checked by (i) carrying out multiple-pass runs, or (ii) for single-pass runs, keeping the process temperature below $T_{c,AGG}$.

LIMITATIONS IN REFINEMENT

The effect of FSP on second-phase particle size and shape refinement in any cast metallic alloy system is well established. However, the degree of refinement during FSP is finite; e.g., the Si particles present in the interdendritic region for a cast Al-Si alloy can only be refined to a limited extent. Studies carried out by various researchers clearly confirm this trend [13,25].

In one of the studies, Ma et al. [25] carried out multiple-pass FSP on a cast A356 plate to create a large processed region. A tool rotation rate of 700 rpm and a traverse speed of 203 mm/min were used for this FSP run. A total of five passes were made, with the FSP tool shifted by one-half the pin diameter between each run. The retreating side of the previous pass was processed during subsequent passes. A transverse cross-sectional macrograph of the processed region is shown in Figure 2.20. The size and morphology of the second-phase Si particles

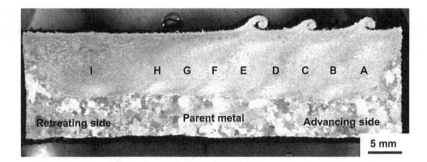

Figure 2.20 Macrograph of transverse cross-section of the processed region [25].

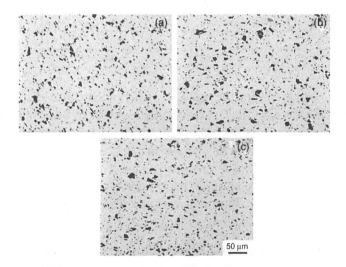

Figure 2.21 Si particle distribution across the process region in a multipass FSPed A356 plate. (a) first pass, (b) second to third transition zone, and (iii) fifth pass (locations A, D, and I in Figure 2.20) [25].

from various locations (marked by A, B, C, D, etc. in Figure 2.20) were determined in detail. A few representative micrographs from various locations are shown in Figure 2.21. Table 2.5 summarizes the size and aspect ratio of Si particles in various regions. The Si particle size and shape data for as-cast and single-pass FSP conditions is also included for comparison. It can be readily seen that increasing the number of passes did not result in further breakdown of Si particle size and shape. The size of Si particles remained rather uniform across the entire process region, and is comparable to Si particle size and shape obtained in a single-pass FSP run. In a separate study, Cui et al. [13] carried out multiple-pass FSP on a cast A356 plate. Tool rotation rate of 1500 rpm and tool travel speed of 300 mm/min were used in this study. A total of

Table 2.5 Size and Aspect Ratio of Si Particles in an A356 Plate Undergoing Five-Pass FSP [25]			
Processing Condition	Location	Particle Size (μm)	Aspect Ratio
Fifth pass FSP	Point A, first pass	2.47 ± 1.89	1.96 ± 0.92
	Point B, first- to second-pass transition	2.56 ± 2.17	1.84 ± 0.74
	Point C, second pass	2.52 ± 2.08	1.85 ± 0.76
	Point D, second- to third-pass transition	2.43 ± 1.89	1.90 ± 0.86
	Point E, third pass	2.54 ± 1.97	1.83 ± 0.77
	Point F, third- to fourth-pass transition	2.40 ± 1.88	1.90 ± 0.81
	Point G, fourth pass	2.38 ± 1.82	1.85 ± 0.77
	Point H, fourth- to fifth-pass transition	2.24 ± 1.69	1.85 ± 0.75
	Point I, fifth pass	2.37 ± 1.89	1.85 ± 0.74
Cast		16.75 ± 9.21	5.92 ± 4.34
Single-pass FSP	Nugget center	2.48 ± 2.02	1.94 ± 0.88

five passes were made for the multiple-pass run. Size and shape analysis of Si particles revealed that increasing the number of passes has a minimal effect on particle refinement. Average Si particle size in the as-cast condition was 12.91 μm. The average Si particle size at the end of the first pass was 3.14 μm and after the fifth pass it was 2.97 μm. The particle size analysis clearly showed that highest level of refinement was attained in the first pass. During subsequent passes, the volume fraction of finer particles increased only slightly.

During FSP, SPD takes place inside the SZ, leading to repeated particle fracture and refinement. However, the particle fracture strength is inversely proportional to the particle size [26]. Therefore, the particle refinement effect during FSP becomes limited once the average particle size reaches a threshold value, e.g., $\sim 2-3$ μm in a cast Al-Si alloy system. This, of course, would depend on the alloy system and the nature of the second-phase particles.

REFERENCES

[1] Thomas WM, et al. Friction stir butt welding, International Patent Application PCT/GB92/02203 and G. B. Patent Application 9125978.8; Dec 1991.

[2] Mishra RS. Friction stir welding and processing. Materials Park, Ohio: ASM International; 2007.

[3] Mishra RS, Mahoney MW, McFadden SX, Mara NA, Mukherjee AK. Scripta Mater 2000;42:163−8.

[4] Ma ZY, Mishra RS, Mahoney MW. Acta Mater 2002;50:4419−30.

[5] Ma ZY. Metall Mater Trans A 2008;39A:642−58.

[6] Ma ZY, Sharma SR, Mishra RS. Mat Sci Eng A 2006;433:269−78.

[7] Krishnan KN. Mat Sci Eng A 2002;A327:246−51.

[8] Ma ZY, Sharma SR, Mishra RS. Metall Mater Trans A 2006;37A:3323−36.

[9] Santella ML, Engstrom T, Storjohann D, Pan TY. Scripta Mater 2005;53:201−6.

[10] Lee WB, Yeon YM, Jung SB. Mat Sci Eng A 2003;A355:154−9.

[11] Jana S, Mishra RS, Baumann JA, Grant GJ. Metall Mater Trans A 2010;41A:2507−21.

[12] Alidokht SA, Abdollah-zadeh A, Soleymani S, Saeid T, Assadi H. Mater Charact 2012;63:90−7.

[13] Cui GR, Ni DR, Ma ZY, Li SX. Metall Mater Trans A 2014;45A:5318−31.

[14] Humphreys FJ, Hatherly M. Recrystallization and related annealing phenomena. 2nd ed. Tarrytown, New York: Pergamon Elsevier Science Inc.; 2002.

[15] Su J-Q, Nelson TW, Sterling CJ. Scripta Mater 2005;52:135−40.

[16] Jata KV, Semiatin SL. Scripta Mater 2000;43:743−9.

[17] Su J-Q, Nelson TW, Mishra R, Mahoney M. Acta Mater 2003;51:713−29.

[18] Fonda RW, Bingert JF, Colligan KJ. Scripta Mater 2004;51:243−8.

[19] Prangnell PB, Heason CP. Acta Mater 2005;53:3179−92.

[20] Rhodes CG, Mahoney MW, Bingel WH, Calabrese M. Scripta Mater 2003;48:1451−5.

[21] Su J-Q, Nelson TW, Sterling CJ. J Mater Res 2003;18:1757−60.

[22] Su J-Q, Nelson TW, Sterling CJ. Philos Mag 2006;86:1−24.

[23] Jana S, Mishra RS, Baumann JA, Grant G. Mat Sci Eng A 2010;A528:189−99.

[24] Jana S, Grant, GJ. Unpublished work.

[25] Ma ZY, Sharma SR, Mishra RS. Scripta Mater 2006;54:1623−6.

[26] Hall JN, Jones JW, Sachdev AK. Mat Sci Eng A 1994;A183:69−80.

CHAPTER 3

Mechanical Properties Enhancement

In Chapter 2, the effect of friction stir processing (FSP) on microstructural refinement in cast metallic systems was discussed. Microstructural modification and refinement leads to significant enhancement in both quasistatic (tensile) and dynamic (fatigue) properties. The greatest impact of FSP is on tensile ductility. Elimination of porosity, refinement of second-phase particles, removal of dendritic structure, etc. lead to dramatic improvement in tensile ductility in cast alloys. In general, there is also improvement in yield strength and tensile strength after FSP. In the following sections, the effects of FSP on mechanical properties in different cast alloy systems are discussed in more detail.

QUASISTATIC PROPERTIES

Aluminum Alloys

FSP has been mostly applied to various cast Al-Si alloys, which are a very important class of materials because they are used extensively in different automotive components. The effect of FSP on A356 alloy tensile properties was studied in detail by Ma et al. [1]. Mini-tensile specimens were obtained from a processed nugget for determining the mechanical properties at room temperature. The effect of post-processing heat treatment was also studied in this work. Tensile properties are summarized in Tables 3.1 and 3.2. The beneficial effect of FSP on tensile elongation is apparent from the data. Tensile ductility improved by more than one order of magnitude as a result of FSP. Additionally, FSP results in improvement of yield strength and tensile strength. However, change in strength values is associated with the FSP parameters used.

A356, which is a precipitation-hardenable class of alloy, gains its strength from homogeneous precipitation of very fine Mg_2Si phases. During the FSP, coarse Mg_2Si phases present in the cast plate go into

Friction Stir Casting Modification for Enhanced Structural Efficiency. DOI: http://dx.doi.org/10.1016/B978-0-12-803359-3.00003-0

Table 3.1 Room Temperature Tensile Properties of As-Cast and FSPed A356; Strain Rate = 1×10^{-3} s^{-1} [1]

Materials	As-Cast or As-FSP			Aged (155 °C/4 h)		
	UTS (MPa)	YS (MPa)	Elongation (Pct)	UTS (MPa)	YS (MPa)	Elongation (Pct)
As-cast	169 ± 8	132 ± 3	3 ± 1	153 ± 6	138 ± 4	2 ± 1
FSP, 300 rpm−51 mm/min (S-pin)	205 ± 5	134 ± 4	31 ± 2	206 ± 5	137 ± 7	29 ± 2
FSP, 700 rpm−203 mm/min (S-pin)	242 ± 4	149 ± 7	31 ± 0	247 ± 6	169 ± 7	28 ± 1
FSP, 900 rpm−203 mm/min (S-pin)	264 ± 3	168 ± 9	31 ± 2	288 ± 4	228 ± 7	25 ± 2
FSP, 1100 rpm−203 mm/min (S-pin)	242 ± 3	157 ± 2	33 ± 1	265 ± 2	205 ± 5	23 ± 5
FSP, 300 rpm−51 mm/min (T-pin)	202 ± 4	137 ± 4	30 ± 1	212 ± 5	153 ± 16	26 ± 2
FSP, 700 rpm−203 mm/min (T-pin)	251 ± 4	171 ± 12	31 ± 1	281 ± 4	209 ± 2	26 ± 2
FSP, 900 rpm−203 mm/min (T-pin)	232 ± 3	140 ± 7	38 ± 1	275 ± 3	204 ± 5	30 ± 1
FSP, 900 rpm−203 mm/min (T-pin)-2 pass	255 ± 1	162 ± 5	34 ± 1	304 ± 3	236 ± 4	25 ± 1
FSP, 1100 rpm−203 mm/min (T-pin)	247 ± 2	155 ± 6	35 ± 1	256 ± 2	177 ± 6	31 ± 4
FSP, 300 rpm−51 mm/min (C-pin)	178 ± 2	124 ± 4	31 ± 3	175 ± 1	119 ± 5	32 ± 1
FSP, 700 rpm−203 mm/min (C-pin)	256 ± 4	169 ± 3	28 ± 1	264 ± 3	203 ± 7	21 ± 1

Table 3.2 Room Temperature Tensile Properties of T6-Treated Cast and FSPed A356; Strain Rate = 1×10^{-3} s^{-1} [1]

Materials	T6 (540 °C/4 h + 155 °C/4 h)		
	UTS (MPa)	YS (MPa)	Elongation (Pct)
As-cast	220 ± 10	210 ± 8	2 ± 1
FSP, 300 rpm−51 mm/min (T-pin)	307 ± 12	232 ± 12	20 ± 1
FSP, 700 rpm−203 mm/min (T-pin)	301 ± 6	216 ± 11	28 ± 2
FSP, 900 rpm−203 mm/min (T-pin)	297 ± 8	213 ± 5	30 ± 2
FSP, 900 rpm−203 mm/min (T-pin)-two pass	292 ± 27	207 ± 24	28 ± 9
FSP, 1100 rpm−203 mm/min (T-pin)	295 ± 6	212 ± 5	28 ± 3

Al solid solution through dissolution because of the imposed process heat and deformation. The intense deformation during the FSP results in much faster dissolution of Mg_2Si phases compared to a regular solution treatment step. However, the volume fraction of Mg_2Si phases that can go into solid solution is intimately linked with peak process temperature and postprocess cooling rate. The particular combination of FSP parameters that could dissolve and retain a higher volume fraction of Mg_2Si phases in Al solid solution would result in better yield strength, especially after aging treatment. This trend is apparent from the data shown in Table 3.1. Part of the data shown in Table 3.1 (T-pin single-pass runs) for which peak process temperature data was available is plotted in Figure 3.1. The highest yield strength resulted from the 700-rpm, 203-mm/min FSP condition. It is likely that the combination of high process temperature and subsequent fast cooling plays a major role in determining which FSP condition produced the best strength properties. The yield strength of the as-cast alloy is indicated by a horizontal line on this plot for reference.

After T6 treatment, yield strength and tensile strength of both the as-cast and FSP material increase significantly, whereas ductility decreases compared to the as-FSPed condition. However, the difference in yield strength between the as-cast and the FSPed conditions is absent after T6 treatment. A solution treatment step during T6 treatment leads to complete dissolution of Mg_2Si phases available in the

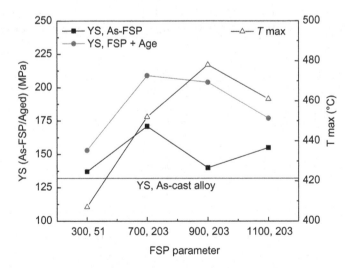

Figure 3.1 Relation between yield strength (YS) and process parameters [1].

A356 system. As a result, the volume fraction of fine Mg_2Si precipitates responsible for strength in the A356 system becomes similar in both the as-cast and the FSP conditions, and therefore the yield strength numbers are similar.

According to Ma et al. [1], the low ductility associated with the as-cast A356 alloy could be attributed to presence of (i) casting porosity and (ii) long acicular Si particles in the interdendritic region. Both the factors led to locally high stress concentration and easy crack nucleation during tensile testing at either Si-particle/Al-matrix interfaces or at casting defects. Cracks, once initiated, find easy growth paths through interdendritic regions, and ultimately result in low ductility and tensile strength. During FSP, the rotating tool breaks down the needle-like Si particles, closes porosity, and homogenizes the microstructure. The presence of refined and spheroidized Si particles, which are more resistant to fracture, and closure of casting porosity help prevent easy void nucleation during tensile tests. As a result, significant improvement in tensile ductility is observed. In fact, the level of ductility enhancement is found to vary with FSP parameters. Higher tool rotation rate, in general, leads to higher ductility, as it results in a higher degree of Si particle refinement.

In a separate study, Jana et al. [2] determined the effects of FSP on tensile properties for a cast F357 alloy. A process parameter optimization study, which included both the tool rotation rate and tool traverse speed, was conducted in this work. Figure 3.2 compares yield and tensile strength in the as-FSPed condition and the as-cast condition.

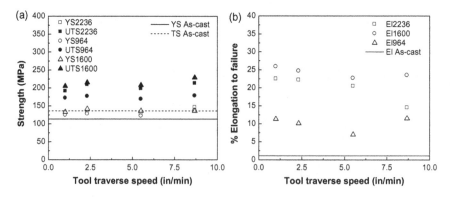

Figure 3.2 Comparison of tensile properties of as-FSPed and as-cast conditions of an F357 alloy: (a) strength and (b) ductility [2].

The yield strength of as-cast material is ~113 MPa and that of as-FSPed materials is slightly higher. As all processing combinations result in similar yield strength values of ~135 MPa, no relationship was observed between yield strength and processing conditions. FSP did significantly improve tensile strength. At 2236 and 1600 rpm tool rotation rates, FSPed material had a tensile strength of ~200 MPa, whereas the tensile strength of the as-cast alloy was 136 MPa. A tool rotation rate of 964 rpm combined with various traverse speeds produced the lowest tensile strength values. However, the most significant effect of FSP is on ductility, as shown in Figure 3.2b. As a result of FSP, elongation to failure increased by close to one order of magnitude. The elongation to failure for the as-cast material was ~1.1%, whereas at the higher tool rotation rates, elongation was consistently greater than 20%. Of all process parameter combinations, those using a 964-rpm tool rotation rate produced the lowest ductility values (<12%). The effectiveness of using higher tool rotation rates in order to achieve higher tensile elongation is confirmed by Jana et al. as well [2].

Tensile properties of T6-treated FSP and cast samples are shown in Figure 3.3. The T6 treatment led to significant strength improvement. As-cast material had a yield strength of 290 MPa, a tensile strength of 300 MPa, and elongation to failure of 0.7%. As shown in Figure 3.3a, FSPed material had similar yield strength values in most conditions, although these values were slightly lower for the lower tool traverse speeds at 964 and 1600 rpm tool rotation rates. FSP significantly increased the tensile strength in the T6 condition as well. Further, a tool rotation rate of 2236 rpm generally produced a better tensile

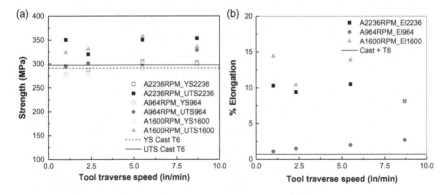

Figure 3.3 Comparison of tensile properties of FSP + T6 and cast + T6 F357 alloy: (a) strength and (b) ductility [2].

strength than other conditions. The highest tensile strength of ~350 MPa was recorded for a 5.5-in/min tool traverse speed at tool rotation rates of 1600 and 2236 rpm. As shown in Figure 3.3b, a tool rotation rate of 964 rpm at all tool traverse speeds produced the lowest ductility. For 1600 and 2236 rpm tool rotation rates, ductility increased by at least one order of magnitude over the cast + T6 material. Further, the highest tool traverse speed of 8.68 in/min resulted in lower ductility than other conditions. The process parameter combination of 2236 rpm and 5.5 in/min resulted in the best tensile properties in T6-treated condition.

Nakata et al. [3] studied the effect of FSP on tensile behavior in a die-cast Al alloy, ADC12 (Al-2.3Cu-11.8Si-0.2Mg-0.5Zn, wt%). Multiple-pass FSP runs were carried out to cover a wide region of the cast plate. Formation of cold flakes in a die-cast Al alloy leads to degradation of tensile properties, tensile ductility in particular. It was shown that FSP was able to (i) eliminate cold flake defects, (ii) disperse Si particles uniformly, and (iii) refine Al-matrix grain size. As a result, tensile strength improved by a factor of 1.7, while elongation improved by a factor of 3.5 over the base material.

A limited study of FSP on cast Al-Cu alloys (2XX series) has been conducted as well [4,5]. This system does not contain large volume fractions of second-phase Si particles. However, small volume fractions of constituent phases, Cu_2FeAl_7, and significant porosity are found in this class of alloy. The large volume fraction of porosity is partly attributed to low castability of this alloy. Therefore, use of this class of alloy is limited, although it can offer higher strength, especially at elevated temperature. Kapoor et al. [4] carried out FSP on cast A206 alloy, and subsequently subjected the FSP specimens to various heat treatment schedules. FSP led to (i) elimination of casting pores, (ii) refined intermetallic particles, and (iii) refined matrix grain size. Tensile test was carried out at room temperature at a strain rate of $1 \times 10^{-3} \, s^{-1}$. Yield strength in cast + T4 condition was 270 MPa, and tensile strength was 310 MPa, with 3% tensile ductility. FSP led to a significant increase in tensile strength from 310 to 405 MPa, and ductility from 3% to 21%. However, yield strength remained the same. The FSP + T4 condition showed similar strength values. T6 treatment on FSP samples resulted in an increase of yield strength, while tensile strength remained similar and ductility dropped. A T7 treatment led to

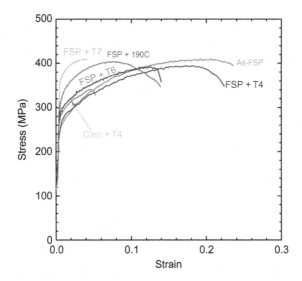

Figure 3.4 Tensile stress–strain plots for A206 alloy: effect of FSP and heat treatment [4].

further increase in yield strength and associated reduction in ductility. The results of the tensile tests are summarized in Figure 3.4. The effects of various heat treatment schedules are captured graphically in Figure 3.5. Gradual increase in yield strength and concurrent drop in ductility is apparent due to the change in aging parameters.

Magnesium Alloys

Magnesium alloys are an important class of lightweight structural materials because of their low density and good mechanical strength. However, the limited number of available slip systems in hexagonal closed packed (hcp) crystal structure results in poor workability, especially in wrought Mg alloys. Grain refinement is an effective technique to address poor workability in Mg alloys. It also leads to enhancement in mechanical properties through the Hall–Petch effect. Many researchers applied FSP to various wrought Mg alloys to generate fine grain structure.

Feng and Ma [6] applied FSP to AZ91 (Mg-8.5Al-0.5Zn, wt%), a common Mg alloy used for producing die-cast components. The microstructure of as-cast AZ91 consists of Mg dendrites, an Al-rich solid solution, and a network of eutectic β-$Mg_{17}Al_{12}$ intermetallic particles in the interdendritic region. The presence of a continuous network of β particles results in poor ductility and mechanical properties.

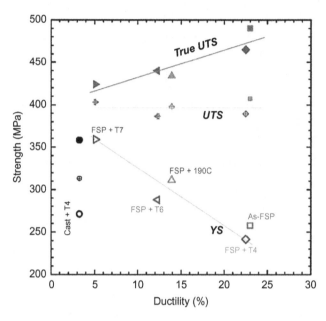

Figure 3.5 Yield strength and ultimate tensile strength (UTS) as a function of ductility; role of various heat treatments in increasing yield strength is apparent [4].

To improve the mechanical properties, this continuous network of intermetallic particles needs to be broken, refined, and distributed evenly inside the Mg matrix. Feng and Ma showed that FSP could be applied successfully to cast Mg alloys to improve their properties. FSP was carried out at a tool rotation rate of 400 rpm and a traverse speed of 100 mm/min. A two-pass run with 100% nugget overlap was employed in order to achieve better β particle refinement and distribution. X-ray diffraction (XRD) together with energy dispersive spectroscopy analysis indicated that FSP led to significant dissolution of $Mg_{17}Al_{12}$ phase, and subsequently the Mg matrix became supersaturated with Al due to β phase dissolution and rapid cooling. Refinement of the as-cast structure as a result of FSP is evident in Figure 3.6. The average grain size inside the FSP nugget was found to be ~15 μm. It is worth mentioning that in Mg-Al alloys, complete dissolution of $Mg_{17}Al_{12}$ can take up to 40 h during solution treatment at 410 °C. However, during FSP, a much faster dissolution of $Mg_{17}Al_{12}$ phases occur inside the nugget due to severe plastic deformation and associated high strain and strain rate. Mechanical properties of the studied AZ91 alloy in as-cast versus FSP conditions are summarized in Figure 3.7. An aging treatment was carried out at 168 °C for 16 h on a

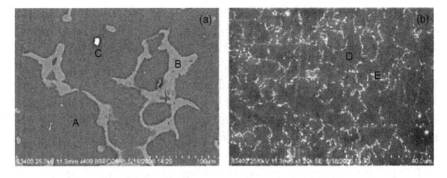

Figure 3.6 Microstructure of AZ91 alloy: (a) as-cast (b) FSP nugget [6].

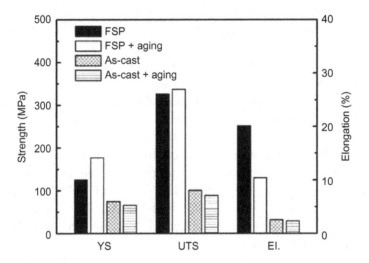

Figure 3.7 Tensile properties of AZ91 alloy [6].

few FSP samples in order to form very fine β-$Mg_{17}Al_{12}$ precipitates throughout the Mg matrix. The as-cast AZ91 alloy showed lower yield strength and tensile strength (73 and 111 MPa, respectively) and a low elongation to failure (2.5%) due to the presence of a continuous network of β-$Mg_{17}Al_{12}$ particles. Considerable improvement of mechanical properties inside the FSP nugget occurs due to (i) breakup and uniform distribution of β-$Mg_{17}Al_{12}$ particles, (ii) significant grain refinement, and (iii) solid solution strengthening by Al.

In a separate study, Feng et al. [7] applied FSP to as-cast AZ80 alloy (Mg-8.5Al-0.5Zn-0.12Mn, wt%). Three different FSP schedules were studied: (i) single pass, (ii) single pass with pre-solution treatment (pre-ST) at 415 °C for 16 h, and (iii) two passes. A tool rotation rate of 400 rpm and a tool traverse speed of 100 mm/min were used in this study. After FSP, the parts were subjected to aging at 177 °C for 10 h. Room temperature tensile test results at a strain rate of $1 \times 10^{-3}\,\text{s}^{-1}$ are summarized in Table 3.3. The as-received AZ80 casting showed lower yield strength and tensile strength (84 and 121 MPa, respectively), and an elongation of 4%. Single-pass FSP resulted in significant strength enhancement. However, the ductility dropped after single-pass FSP, and the reduction in ductility is more pronounced after aging treatment. On the other hand, both pre-ST FSP and two-pass FSP resulted in much improved mechanical properties. Aging treatment, in general, increased yield strength and reduced ductility. Among the three FSP schedules studied, the aged two-pass FSP sample showed the highest tensile strength of 356 MPa, a yield strength of 193 MPa, and an elongation of 17%. Improvement in mechanical properties after two-pass FSP and pre-ST FSP schedules could be linked to β-$Mg_{17}Al_{12}$ dissolution and refinement, homogeneous microstructure, and grain size refinement. Moreover, the two-pass FSP schedule resulted in a smaller grain size together with higher Al concentration inside the Mg matrix, and as a result higher yield strength could be realized after aging treatment due to precipitation of a very fine β phase. In comparison, single-pass FSP was not able to completely eliminate the β-$Mg_{17}Al_{12}$ network, and the ductility value remained poor. Figure 3.8 depicts the effect of various FSP schedules on β-$Mg_{17}Al_{12}$ morphology.

Table 3.3 Tensile Properties of AZ80 Alloy [7]			
Conditions	YS (MPa)	UTS (MPa)	El. (Pct)
As-cast	84.1	120.9	4.1
As-cast + solutionizing	94.3	176.5	8.8
Single-pass FSP	134.6	188.8	3.0
Aged single-pass FSP	156.5	213.3	1.2
Pre-ST FSP	109.2	308.8	25.2
Aged pre-ST FSP	169.2	336.5	16.8
Two-pass FSP	136.7	327.3	25.0
Aged two-pass FSP	192.8	355.7	17.0
YS, yield strength; UTS, ultimate tensile strength; El., elongation.			

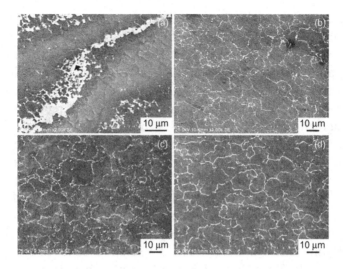

Figure 3.8 β-Mg₁₇Al₁₂ morphology: (a) and (b) single-pass FSP, (c) pre-ST FSP, and (d) two-pass FSP [7].

Table 3.4 Tensile Properties in an HPDC AM60B Alloy after FSP [8]

Condition	0.2% Yield Strength (MPa)	Tensile Strength (MPa)	Total Elongation (%)
1250-rpm Base metal	114	205	9.6
FSP	140	280	22.6
2500-rpm Base metal	119	192	7.2
FSP	141	266	15.7
ASTM B94-05	130	220	8
			(in 50 mm)

Santella et al. [8] applied FSP to a high pressure die-cast (HPDC) AM60B alloy (Mg-6Al-0.13Mn, wt%) to modify the surface. FSP runs were made at two different tool rotation rates, 1250 and 2500 rpm, and at a tool travel speed of 102 mm/min. Tensile properties from the processed region are summarized in Table 3.4. FSP led to modest enhancement in strength and significant increase in ductility. Elimination of the coarse β-Mg₁₇Al₁₂ network together with significant grain refinement was found to be the reason for mechanical property improvement.

Rare-earth (RE)-element-containing Mg alloys are another important class of materials because of their excellent high-temperature strength and creep resistance. The effect of FSP in RE-containing alloys has been examined as well. Tsujikawa et al. [9] investigated the effect of FSP in a Mg-Y-Zn cast alloy, and concluded that formation of ultrafine grain structure inside the processed nugget leads to exceptional strengthening. Freeny et al. [10] carried out FSP on a high-strength and creep-resistant cast WE43 alloy (Mg-4.0Y-2.25Nd-1.0RE-0.6Zr, wt%). It was reported that variation in FSP thermal input played a critical role in determining dissolution and reprecipitation of the strengthening β-Mg-Y-Nd phases, but only had a weak effect on grain size. It was further concluded that the best tensile properties are obtained when a solution treatment step is carried out prior to FSP, followed by an aging treatment. For the best processing condition, a tensile strength of 303 MPa, a yield strength of 253 MPa, and an elongation of 17% were obtained, representing 125% improvement over the cast + T6 condition. In another study, Freeny and Mishra [11] carried out FSP on Elektron 21 (ASTM EV31A), a sand cast Mg alloy. EV31A (Mg-0.5Zn-3.1Nd-1.7Gd-0.3Zr, wt%) has better castability and provides a low-cost alternative to the high-strength, creep-resistant Mg-Y Elektron WE43 alloy. A tool rotation rate of 400 rpm and a traverse speed of 102 mm/min were used. A T6 treatment, which included solution treatment at 520 °C for 8 h, followed by quenching and subsequent aging at 200 °C for 16 h, was carried out to maximize strength. Mechanical properties after FSP improved significantly due to a combination of reduction in grain size, breakage of secondary particles, and dissolution of strengthening phases. Grain size could be reduced from 78 to 3 μm after FSP. The results of tensile tests are summarized in Figure 3.9. It can be seen that FSP with subsequent aging of a T6-treated plate resulted in a peak yield strength of 266 MPa, an improvement of about 40% over the as-received T6 plate.

Xiao et al. [12] investigated the effect of FSP on a cast Mg-Gd-Y alloy (GW103) with a nominal composition of Mg-10Gd-3Y-0.5Zr (wt%). This class of RE-containing Mg alloy is especially attractive as it contains high melting point intermetallic $Mg_5(Gd,Y)$, and thus offers improved thermal stability. Cast Mg-Gd-Y alloy contains a continuous network of coarse eutectic β-$Mg_5(Gd,Y)$ particles, and α-Mg grains, and thus are brittle in nature. FSP of as-cast plate was carried out at a tool rotation rate of 800 rpm and at a tool travel speed of 50 mm/min. Subsequent aging treatment was carried out at 225 °C for 13 h. The microstructure of the processed plate is shown in Figure 3.10. The average grain size in the as-cast condition was 145 μm. FSP led to tremendous grain refinement—the resulting average grain size was 6 μm.

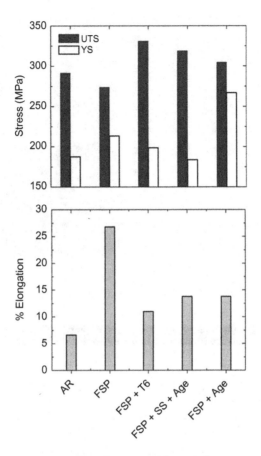

Figure 3.9 Average room temperature tensile properties in an Elektron 21 alloy, as-received (AR) and after FSP and subsequent heat treatment schedules, strain rate $1 \times 10^{-3} s^{-1}$ [11].

Additionally, no β-Mg$_5$(Gd,Y) could be detected inside the FSP stir zone, even at high magnification. The same result was confirmed by XRD through lattice parameter shift analysis. Dissolution of Mg-Gd phase through conventional heat treatment methods requires several hours of exposure at a very high temperature of $\sim$500 °C. Therefore, it is apparent that severe plastic deformation during FSP aids in fast dissolution of β-Mg$_5$(Gd,Y) particles through accelerated diffusion rate and shorter diffusion distance. Subsequent aging treatment after FSP led to reprecipitation of very fine ($\sim$10 nm) metastable β'' and β' phases throughout the grain-refined α-Mg matrix. Results of room temperature tensile tests done at a strain rate of 1×10^{-3} s^{-1} are summarized in Table 3.5. FSP leads to moderate increase in yield strength, while ductility improves considerably because of the elimination of the

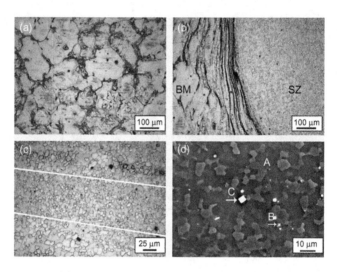

Figure 3.10 Microstructure of Mg-Gd-Y alloy: (a) Base metal (BM), (b) (SZ) and BM interface, (c) inside SZ, (d) SZ at higher magnification [12].

Table 3.5 Tensile Properties of Mg-Gd-Y Alloy Under Various Conditions [12]			
Sample	**UTS (MPa)**	**YS (MPa)**	**El. (%)**
AS-cast	187	178	3.2
FSP	312	210	19
FSP + aging	439	330	3.4

coarse eutectic β-Mg$_5$(Gd,Y) network. However, aging after FSP leads to a significant drop in ductility, while strength values increase. It is believed that the formation of a particular combination of strengthening β'' and β' phases could be causing such a drop in plasticity. Optimization of FSP parameters can help in overcoming this particular challenge.

In a later study, Yang et al. [13] employed three different travel speeds (25, 50, and 100 mm/min) at a tool rotation rate of 800 rpm for FSP of a cast Mg-Gd-Y alloy (GW103). It was shown that the ductility of FSP samples increased with increase in tool travel speed, which could be linked to a slight decrease in average grain size. However, the strength values did not change as a function of travel speed. Aging after FSP (T5 conditions) resulted in increase in yield strength and tensile strength. On the other hand, a drop in ductility occurred with aging. The tensile test results are summarized in Tables 3.6 and 3.7.

Table 3.6 Tensile Properties of Cast Mg-Gd-Y Alloy Under Various FSP Conditions [13]			
Temper	YS (MPa)	UTS (MPa)	El. (pct)
Cast-T4	157 ± 6.3	224 ± 6.3	14.4 ± 2.6
FSP, 800 rpm-25 mm/min	268 ± 2.2	357 ± 0.6	24.0 ± 1.5
FSP, 800 rpm-50 mm/min	281 ± 6.6	363 ± 2.3	29.1 ± 0.4
FSP, 800 rpm-100 mm/min	214 ± 6.6	322 ± 2.1	31.0 ± 1.5

Table 3.7 Tensile Properties of FSP Sample (800 rpm, 25 mm/min) in the Transverse (TD) and Longitudinal (LD) Directions [13]				
Specimen	Orientation	YS (MPa)	UTS (MPa)	El. (pct)
FSP	TD	268 ± 2.2	357 ± 0.6	24.0 ± 1.5
	LD	223 ± 3.4	323 ± 7.2	18.8 ± 0.7
FSP-T5	TD	327 ± 4.5	424 ± 8.8	6.0 ± 1.3
	LD	329 ± 4.9	418 ± 7.1	3.4 ± 0.3

Titanium Alloys

A limited amount of study has been done to understand the influence of FSP in Ti-alloy castings. Pilchak and Williams [14] reported a detailed study of FSP carried on investment-cast Ti6-4 alloy. In order to understand the role of microstructure on mechanical properties, FSP was carried out above and below the β transus (the lowest temperature at which a 100% β phase can exist) to create various ratios of α and β phases in the matrix. Details of the FSP parameters used in this study are summarized in Table 3.8. The α/β and high-α/β conditions are created by maintaining FSP process temperature below the β transus, while the β transus is exceeded to create the β FSP condition. During sub-transus processing, the lamellar α structure becomes spheroidized due to intense deformation during FSP, but cannot coarsen due to the huge composition difference between α and adjacent β grains, and thus remains very small in size. Processing above the β transus results in transformation of α/β microstructure into complete β, which later transforms into lamellar α during cooling. As FSP leads to significant β phase refinement (from 1 to 2 mm in as-cast condition to 25 μm), much finer lamellar α structure results.

Tensile test results of the cast and hot isostatically pressed (HIPed) Ti6-4 alloy is summarized in Table 3.9. Average yield strength, UTS, and percent elongation for the as-cast alloy were measured at

Table 3.8 Process Parameters Used for FSP of As-Cast and HIPed Ti6-4 Alloy [14]		
Condition	Tool Travel, cm/min (in/min)	Tool Rotation Speed (rpm)
α/β	10.2 (4)	100
High α/β	10.2 (4)	150
β	10.2 (4)	150

Table 3.9 Tensile Properties of As-Cast Ti6-4 Alloy [14]					
Specimen ID	E (GPa)	Yield Stress (MPa)	Ultimate Tensile Strength (MPa)	Reduction of Area (pct)	ε_f
1	116	823	889	23.2	0.153
2	116	801	853	28.9	0.131
3	116	777	861	28.4	0.163
4	128	796	871	26.4	0.138
Average	–	799	869	26.7	0.146

799 MPa, 869 MPa, and 14.6, respectively. To measure FSP properties, microtensile tests were carried out. A summary of the test results is shown in Table 3.10. The β FSP condition showed the lowest yield strength and UTS, which could be related to the highest effective slip length. The α/β and high-α/β conditions showed an average yield strength of 1091 and 1132 MPa, respectively. Elongation to failure for the β-FSP condition was ~21%, while for both the α/β conditions ~17% was reported. In general, FSP led to refinement of the as-cast microstructure in the Ti6-4 alloy and subsequent strength enhancement, without a deleterious effect on ductility.

FATIGUE PROPERTIES

Mechanical fatigue failure of components takes place under the application of repeated cyclic stress or strain. Fatigue is a time-dependent failure process, and can be divided into three stages: (i) crack initiation, (ii) crack growth, and (iii) ultimate ductile failure. In an effectively defect-free specimen, fatigue crack initiation takes place at the surface. Development of extrusion or intrusion structures at specimen surface through persistent slip band (PSB) formation gives rise to the start of a fatigue crack. Once initiated, fatigue cracks would grow along the planes of easy slip. Once it is well defined, fatigue crack growth (FCG) occurs in a direction normal to maximum tensile stress

Table 3.10 Tensile Properties After FSP of Ti6-4 Alloy [14]				
Specimen ID	Yield Strength (MPa)	Ultimate Tensile Strength (MPa)	Stress at Slope Change (MPa)	ε_f
β Friction stir processed condition				
1	1063	1116	600	0.23
2	1028	1110	557	0.17
3	1041	1085	609	0.22
4	1026	1078	593	0.21
5	1057	1115	592	0.20
6	1053	1105	620	0.21
β FSP average	1045 ± 15.5	1102 ± 16.1	595 ± 21.5	0.21 ± 0.2
High α/β friction stir processed condition				
1	1122	1185	519	0.16
2	1119	1180	563	0.17
3	1154	1198	568	0.18
4	1139	1200	572	0.18
5	1128	1185	660	0.16
High α/β FSP average	1132 ± 14.3	1190 ± 8.9	576 ± 51.4	0.17 ± 0.01
α/β Friction stir processed condition				
1	1012	1059	584	0.14
2	1097	1142	543	0.16
3	1124	1152	630	0.17
4	1119	1158	608	0.17
5	1122	1153	660	0.19
6	1070	1119	592	0.18
α/β FSP average	1091 ± 43.7	1131 ± 37.7	603 ± 40.2	0.17 ± 0.02

in a transgranular manner. Final failure occurs when the crack has reached sufficient length that the remaining section cannot support the applied load. The overall fatigue life of a component is composed of number of stress/strain cycles spent in each stage.

Fatigue life of castings, in particular Al-alloy-based ones, is governed by the size of microstructural defects. Fatigue strength of a casting decreases with increase in defect size. As summarized by Wang et al. [15] in a recent review on this topic, pores and oxide films are the two major sources of defects that are responsible for fatigue failure in aluminum castings. Pores can be either gas-driven or shrinkage driven. Gas-driven pores are typically spherical, while shrinkage-driven pores have a tortuous, convoluted morphology. Oxide films are generated during mold filling, and can be thick or thin depending on the mode of formation. The shape and size of these defects would determine the stress intensity factor during fatigue loading. In the absence of pores and oxides, fatigue cracks initiate on second-phase particles or from PSBs. Further, the largest defect size plays the greatest role in determining fatigue life of a casting, rather than the overall volume fraction. A significant amount of research has been carried out to understand the roles of the shape and size of casting defects (pores, oxides) on fatigue properties in various Al castings. It is generally believed that the shape of the defect has minimal influence on fatigue properties, but the defect size plays a critical role in determining fatigue behavior. As shown in Figure 3.11, a logarithmic plot of fatigue life versus defect size, the trend of achieving enhanced fatigue life with reduction in pore size is obvious. In Figure 3.12, the relative importance of two defects (e.g., pores and oxide films) on fatigue life is illustrated. Pores are found to be more detrimental than oxide films. Further, the location of defects in a casting is important as well. Surface and near-surface defects are more potent than subsurface defects for comparable sizes. Therefore, from a microstructural point of view, FSP appears to be ideal for improving fatigue life/strength of castings: through FSP, (i) porosity can be closed or nearly eliminated, (ii) significant refinement of second phases can be achieved, and (iii) surface or near-surface defects can be easily closed with proper FSP tool design.

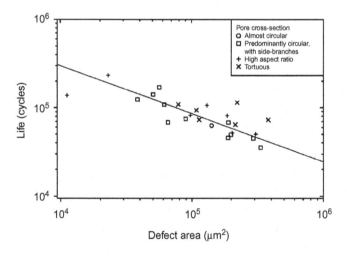

Figure 3.11 The effect of pore size and shape on fatigue life in an Al-7Si-0.6 Mg casting [15].

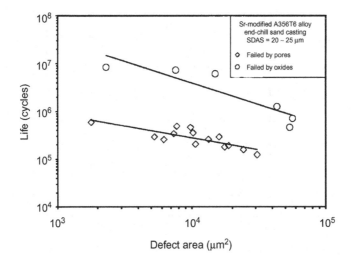

Figure 3.12 The effect of defect size on fatigue life in an A356-T6 casting. The detrimental role of casting porosity is clear [15].

Removal of potential defects from casting through FSP at fatigue-critical locations would help bring out the intrinsic fatigue strength of a cast alloy.

Fatigue properties of Al alloy castings after FSP have been investigated by several researchers. The effect of FSP on the fatigue properties of cast A356 alloy was first reported by Sharma et al. [16]. For fatigue testing, the following two FSP parameter sets were selected: (i) 900 rpm, 8 in/min, standard FSW pin, (ii) 700 rpm, 8 in/min, triflute FSW pin. Fatigue tests were carried out on a closed-loop servo-controlled hydraulic system under axial load at a stress ratio of $R = 0.1$. The frequency was set at 80 Hz. Fatigue test results are shown in Figure 3.13. Fatigue strength or endurance limit increased by more than 80% compared to the as-cast alloy as a result of FSP. Table 3.11 summarizes the results of microstructural analysis carried out. The beneficial effects of FSP on (i) porosity removal and (ii) second-phase particle refinement are clear, having produced such a significant fatigue strength enhancement. The presence of larger Si particles in the as-cast microstructure accelerated crack nucleation. With reduction in particle size as a result of FSP, the probability of crack nucleation was significantly reduced. In the FSP microstructure, significant refinement together with homogeneous distribution of Si particles and lower aspect ratio led to increased plastic deformation during cyclic crack-tip

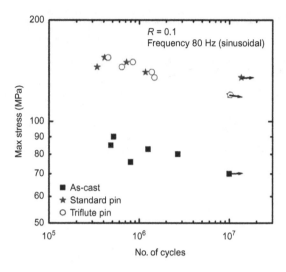

Figure 3.13 S−N plot for A356 alloy [16].

Table 3.11 Influence of FSP on Particle Size and Porosity Volume Fraction in Cast A356 [16]		
Material	**Particle Area (μm^2)**	**Porosity Volume Fraction (%)**
As-cast 3/5″	69 ± 152	0.95
900 rpm and 203.2 mm/min (standard-pin)	20 ± 26	0.027
700 rpm and 203.2 mm/min (triflute-pin)	16.5 ± 26.5	0.024

propagation. Therefore, increased crack growth resistance could be realized through increase in crack energy dissipation. No effect of tool geometry on fatigue properties could be identified. Fractography results indicated crack nucleation at casting defects for as-cast samples. The crack path was rough and uneven, showing preference for crack growth along the defects present. For FSPed samples, the fracture surfaces were almost perpendicular to the loading axis. The crack growth was noticed to be along the Si-particle and Al-matrix interface. Crack growth along the particle/matrix interface for FSPed samples caused an increase in effective crack length, or in other words, cracks became more tortuous in nature.

Santella et al. [8] reported the effect of FSP on fatigue properties of A356 and A319 alloys. FSP was carried out at 1000 rpm and 4 in/min. Results from a fully reversed ($R = -1$) fatigue test are summarized in Table 3.12. After FSP, significant extension in fatigue life was

Table 3.12 Comparison of Fatigue Life for Cast and FSPed A356 and A319 [8]			
Condition	Strain Amplitude	Stress (MPa) Amplitude at Half-Life	Number of Cycles to Failure
As-cast 356	0.002	108	7700
FSP 356	0.002	136	93,848
As-cast 319	0.0014	93.6	100,980
FSP 319	0.0014	126	281,442

apparent, and is attributed to (i) reduced porosity, and (ii) refinement and uniform distribution of second-phase particles.

Jana et al. [17] carried out a detailed study to understand the fatigue behavior of cast F357 alloy after FSP. The parameters used for FSP were tool rotation speed 2236 rpm and 5.5 in/min travel speed. Fatigue properties were evaluated in the T6 condition. The as-cast alloy after T6 condition had a yield strength of 290 MPa, an UTS of 300 MPa, and elongation to failure of ~1%. After FSP + T6 treatment, the yield strength was 290 MPa. UTS was 330 MPa and elongation to failure was 15%. Fatigue properties were determined by testing bending fatigue specimens. Because of the large specimen size involved, multiple-pass FSP was carried out to generate a large processed region. Figure 3.14 shows the stress versus cycles to failure (S–N) plot comparing fatigue properties of cast and FSP conditions for a completely reversible bending stress. FSP led to a ~5× enhancement in fatigue life over the as-received cast condition when tested at the same stress level. Additionally, the level of fatigue life enhancement is found to be higher toward lower applied stress levels.

*According to the stress-life approach, the total life of a cyclically loaded component, N_f, consists of (i) the number of cycles required for crack initiation, N_i, and (ii) the number of cycles to propagate dominant fatigue crack to final failure, N_p. N_i can be as low as 0%, for specimens containing defects, to as high as 80% in largely defect-free smooth specimens [18]. Defect-tolerant design approaches, on the other hand, consider crack propagation to be the main event during the entire fatigue life. In this approach, the well-known Paris equation, Eqn (3.1), can be integrated to estimate N_f [18].

$$\frac{da}{dN} = C(\Delta K)^m \tag{3.1}$$

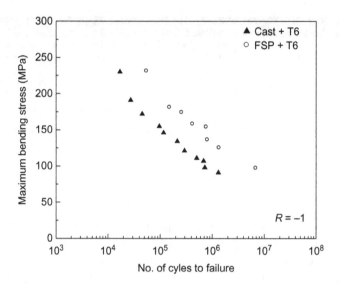

Figure 3.14 S−N plot for F357 alloy [17].

$$N_f = \frac{2}{(m-2)CY^m(\Delta\sigma)^m\pi^{m/2}}\left\{\frac{1}{(a_i)^{(m-2)/2}} - \frac{1}{(a_f)^{(m-2)/2}}\right\} \qquad (3.2)$$

$$N_f = \frac{1}{CY^2(\Delta\sigma)^2\pi}\ln\left(\frac{a_f}{a_i}\right) \qquad (3.3)$$

where da/dN represents crack growth rate, ΔK is the stress intensity factor range, $\Delta\sigma$ is the stress range, m and C are scaling constants, Y is a geometrical factor, a_i is the initial crack size, and a_f is the final crack size. Equation (3.2) is applicable for $m\neq 2$, while Eqn (3.3) is for the $m = 2$ condition. For small defects ($a_i \ll a_f$) and $m \sim 4$, which is the case for A356/357 [19,20], Eqn (3.2) can be further simplified as

$$(\Delta\sigma)^m N_f = \frac{B}{a_i} \qquad (3.4)$$

Wang et al. [15] in their recent review derived a similar relation of the form

$$\sigma_a^m N_f A_i^{(m-2)/4} = B \qquad (3.5)$$

In both Eqns (3.4) and (3.5), B includes various constants. In Eqn (3.5), A_i is the projected defect area normal to the maximum tensile stress. In an effort to correlate fatigue life with initial defect size, Murakami and Endo [21,22] introduced the concept of projected defect area. This was because they noted that the maximum stress intensity factor, K_{Imax}, can be related to the crack area by

$$K_{\text{Imax}} \propto \left(\sqrt{\text{area}}\right)^{1/2} \tag{3.6}$$

Further, it should be mentioned that while deriving Eqn (3.5), the authors assumed the projected defect area to be proportional to the square of the characteristic crack length. Wang et al. [15] showed that the fatigue life in A356/357 is inversely proportional to $\sqrt{A_i}$ by substituting $m = 4$ in Eqn (3.5). Equation (3.4) similarly predicts that fatigue life is inversely proportional to initial defect size.

Postmortem study of the failed specimens showed that porosity in the cast microstructure was responsible for crack nucleation. An example of a failed cast fatigue sample surface (Figure 3.15) shows initiation of cracks from corners of porosity. The fatigue life of the cast alloy is therefore governed by the size of porosity. Additionally, the size of the largest pore rather than the mean pore size is more influential in determining the total fatigue life. No crack initiation was observed with comparatively smaller porosity in the cast fatigue specimens. As FSP results in closure of porosity, fatigue life improves because of reduction in initial defect size. In the case of FSPed microstructure, the cracks initiated at the Si-particle/Al-matrix interface as noted in Figure 3.16. An arrow shows the advancement of such cracks toward both sides from the particle/matrix interface. It was further observed that comparatively bigger Si particles were the sites for crack nucleation in the case

Figure 3.15 Crack initiation at porosity corner in as-cast F357 alloy [17].

Figure 3.16 Crack initiation at Si/Al-matrix interface, FSP + T6 condition, F357 alloy [17].

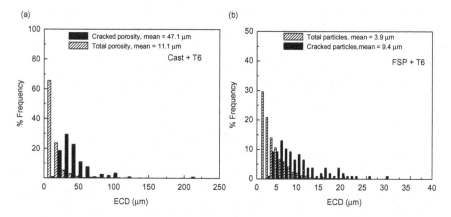

Figure 3.17 Defect size comparisons: (a) cast + T6 versus and (b) FSP + T6. [17]. ECD, equivalent circular diameter.

of FSPed material. The observed fatigue life can be linked to respective initial critical defect size with the assumption that fatigue life is governed by crack growth. The critical defect size in cast fatigue specimens was determined by measuring the size of porosities that showed cracks emanating from corners. In case of the FSPed fatigue specimen, size of comparatively larger Si particles that cracked was determined. The results are shown in Figures 3.17a and b. The statistical average of the size of defects (porosity) responsible for failure in the cast + T6 sample was 47.1 μm. In the FSP + T6 sample, the average size of responsible Si particles was 9.4 μm. The size ratio of mean critical porosity to mean critical Si particles was ~5.0, the same factor by which fatigue life in FSP condition improved over the cast alloy. Although the porosity volume fraction was quite low (~0.2%) in the as-cast condition and mean porosity size was ~10 μm, it was the larger pores that were responsible for early failure in the as-cast condition (largest porosity was 196 μm, i.e., almost 20 × bigger than the average defect size). The effects of FSP are to (i) close down porosity and (ii) refine Si particles, especially the larger particles, which were reduced by more than ~50%. As a result, a much better fatigue response is observed.

FATIGUE CRACK NUCLEATION/GROWTH

A detailed study by Jana et al. [17] investigated crack initiation and crack growth morphology in as-cast versus as-FSPed condition by carrying out postmortem analysis of failed specimens. Fatigue cracks initiate at porosity corners due to the high stress concentration there. Crack growth occurs through both aluminum dendrites as well as interdendritic regions (Figure 3.18). Crack growth paths inside Al dendrites are rather linear. However, a more tortuous pattern is noted when crack interaction with Si particles takes place at interdendritic regions. Both particle/matrix decohesion and particle fracture occur as a consequence. In general, comparatively larger particles fracture and smaller particles show particle/matrix decohesion. The transition from particle decohesion to particle fracture occurs at ~10 μm. Lee et al. [23] compared fatigue crack propagation through microstructures exhibiting fine and coarse Si morphologies in cast Al-Si alloys at the same stress level and stress ratio, and obtained similar results. Since the fracture strength of Si particles is inversely proportional to their size, fine Si particles are much stronger than the critical stress required for particle cracking. Subsequently, fine particles go through decohesion, whereas bigger ones fracture. Further, the propensity for particle cracking increases with higher crack driving force, which increases with increase in crack length at a fixed applied external stress. The study by Jana et al. [17] shows that cracks in the initial stage prefer to

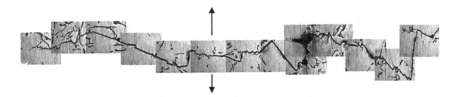

Figure 3.18 Crack profile, cast + T6 condition. Arrows show loading direction [17].

move through the α-Al dendrite rather than interact with Si particles. When the crack is forced to interact with Si particles, particle decohesion occurs, except with bigger and irregularly shaped particles. An example of a big cracked particle is shown by the large arrow in Figure 3.19. Typically particles >10 μm can be termed "big." Changes in a crack growth path can depend on its length. Gall et al. [24] in their work on the fatigue of alloy A356 observed that, near the crack-initiating defect, when the crack length was small, the crack primarily grew straight through the α-Al dendrites, avoiding the eutectic regions. Particle decohesion occurred only when the crack was forced to interact with eutectic regions. As the crack length increased, particle failure morphology changed from decohesion to cracking, and the crack preferred to move mainly through large irregularly shaped Si particles because these regions provided the path of least resistance.

The crack profile in an FSP condition is shown in Figure 3.20. It differs from those in cast conditions in two respects: (i) more crack meandering, and (ii) more crack branching. As FSP leads to elimination of the dendritic structure, Si particles were distributed almost uniformly throughout the α-Al matrix. As a consequence, cracks that initiated at the particle/matrix interface under FSP conditions were forced to interact with Si particles. The mode of particle failure is again governed by particle size at an early crack growth stage when

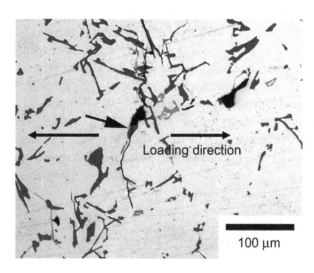

Figure 3.19 Fatigue crack morphology in a cast + T6 F357 alloy. Relatively larger Si particles were fractured in the crack path [17].

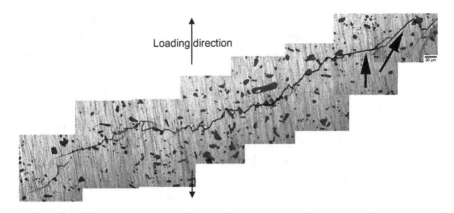

Figure 3.20 Crack profile in FSP + T6 condition. The crack moved along the particle/matrix interface. Failure mode at a particle is either debonding or cleavage, based on particle size. Larger arrows indicate regions of faster crack growth due to lower particle number density. Crack branching is also apparent [17].

the size of the crack is comparatively short. Additionally, it can be inferred from Figure 3.20 that particle debonding is the prominent failure mode because of small particle size; only big particles, whose numbers were limited, fractured. Fatigue cracks experience significant growth retardation when cracks are forced to propagate gradually along the particle/matrix interface [25]. Growth retardation occurs because the effective crack length increases as the crack grows around a network of particles within a ductile matrix. In addition, a tortuous crack path forces the crack tip to grow under unfavorable local mixed-mode loading conditions rather than under a favorable mode I driving force. Changes in the crack growth pattern can also be observed in Figure 3.20. The crack moved in a zigzag manner along the particle/matrix interface for most of its course, but at the right-hand corner of the image, the crack is seen to move in an almost straight line, as indicated by the arrows. Locally, that particular region has a lower number density of Si particles, causing the change in the crack growth pattern. Change in the crack growth rate as a result of alteration of crack interaction mechanism (i.e., growth through Al dendrites and particle fracture in as-cast condition vs particle/Al-matrix decohesion in as-FSPed condition) is the underlying operative mechanism for improved fatigue performance after FSP. Additionally, increased crack branching was also associated with the crack profile in FSP conditions. As noted by Lee et al. [23], this phenomenon leads to a reduction in crack growth by increasing total crack length while reducing the effective stress intensity at the crack tip when the crack branches.

The study by Jana et al. [17] provides further insight into the role of microstructure on fatigue crack initiation/growth morphology. A crack growth study was conducted at $\sigma_{max}/\sigma_y = 0.6$ ($\sigma_{max} = 174$ MPa) for $R = -1$ conditions for both FSPed and cast fatigue samples. The initiation and growth of cracks was recorded optically by stopping the fatigue test intermittently. Total life of the FSP specimen was 2.609×10^5 cycles. Substantial crack nucleation at multiple sites was noted at the end of 20,000 cycles, when testing of the FSP specimen was stopped for the first time (Figures 3.21a and b). Therefore, number of cycles for crack initiation, N_i, represents only 7.6% of the total life, which shows that for the given test conditions, FCG is the major failure mechanism. Although multiple cracks can initiate at various locations, only a few of them become dominant, based on favorable loading conditions. A dominant crack moves along the direction normal to the main loading axis. The dominant crack profile is highlighted in Figure 3.21c with a black arrow indicating the initiation site and a red arrow indicating the location of the crack tip at the end of 20,000 cycles. Subsequent recording of crack tip position after every 20,000 cycles until specimen failure and measuring crack extension, Δa, helps in constructing a fatigue crack growth rate (FCGR) curve. The initial growth direction of cracks is at 45° to the loading axis (Figure 3.21d), indicating stage I or a near-threshold regime of crack growth. As fatigue testing continues, crack nucleation moves from the edge of the test specimen to its interior (Figure 3.21e, crack initiation at Si-particle/Al-matrix interface). Once again, the nature of the crack growth profile depends primarily on the crack length; newly initiated cracks are most often inclined at 45° to the loading direction, whereas old cracks grow at almost 90° to the tensile axis, finally becoming dominant. An example of a dominant crack is shown in Figure 3.21f.

As mentioned previously, the growth rate of fatigue cracks is usually described by Paris' law, Eqn (3.1), which can be further expanded as [18]

$$\frac{da}{dN} = C(\Delta K)^m = C\left(Y\Delta\sigma\sqrt{\pi a}\right)^m = CY^m(\Delta\sigma)^m \pi^{m/2}\left(\sqrt{a}\right)^m \qquad (3.7)$$

where da/dN indicates crack growth increment and ΔK is the stress intensity factor range, which is related to the stress range $\Delta\sigma$, Y is a geometrical factor, a is the crack length, and C and m are scaling constants. For a constant-amplitude fatigue test, da/dN is a function of

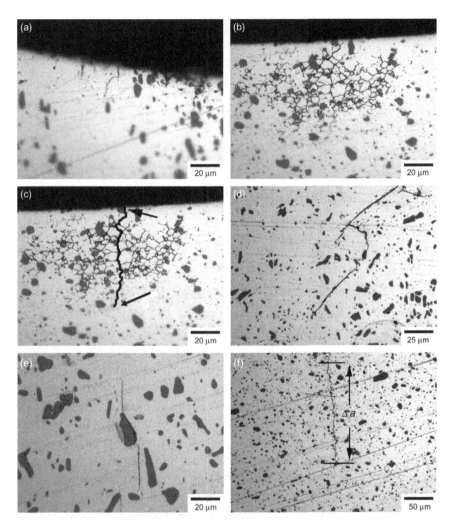

Figure 3.21 Crack initiation and growth in FSP + T6 condition: (a) crack initiating in the α-Al matrix after 20,000 cycles, (b) crack initiation at the particle/matrix interface after 20,000 cycles; crack branching along grain boundaries is also apparent, (c) dominant crack is highlighted (black arrow indicates nucleation site whereas red arrow shows crack tip position after 20,000 cycles), (d) crack showing stage I behavior, (e) cracking at later stages in particle/matrix interface away from specimen edge, and (f) crack extension between two observations; crack tip shown in (c) has grown by Δa [17]. (For interpretation of the references to color in this figure legend and inside the text, the reader is referred to the web version of this book.)

crack length, assuming Y to be a constant. Jana et al. [17] measured crack growth increments from optical images and subsequently determined crack growth rates. Figure 3.22 shows a typical FCGR curve, where average crack growth rate data have been plotted against $\sqrt{a}$ on a log–log scale. The crack growth rate remains a linear function of $\sqrt{a}$

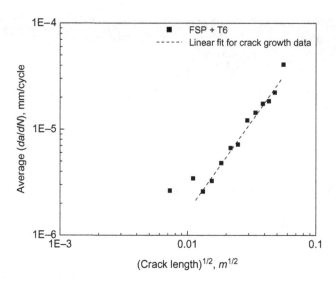

Figure 3.22 Fatigue crack growth rate curve, FSP + T6 condition, F357 alloy [17].

between 6×10^4 and 2.4×10^5, which corresponds to the Paris regime. Below the Paris regime, there is an oscillatory behavior of crack growth rate, whereas above it, there is a precipitous rise in crack growth rate with increasing $\sqrt{a}$. As FCG changed from stage I to stage II (Paris regime) around 60,000 cycles (Figure 3.22), significant changes associated with the crack growth profile was also noticed. During stage I, crack growth occurs predominantly by single shear in the direction of primary slip systems and therefore leads to a zigzag crack path. On the other hand, stage II in general involves duplex slip, resulting in a crack path normal to the far-field tensile axis [18]. Such changes in the crack growth path were observed by Jana et al. [17], as shown in Figures 3.23a−e. The crack, which was almost at a 45° angle to the tensile axis at the end of 40,000 cycles (Figure 3.23a), gradually changed its path. At the end of 80,000 cycles, a slight deviation from the 45° line was apparent. This change becomes more evident at 100,000 cycles, and after 120,000 cycles the same crack was propagating almost perpendicularly to the tensile axis.

The cast fatigue specimen showed a life of 45,500 cycles. Cracks originated at porosities when the test was stopped after the first 5000 cycles, implying a 10% or lower crack initiation period. FCGR curves of both as-cast and as-FSPed material are shown in Figure 3.24. Average crack growth rate is close to one order of magnitude higher in

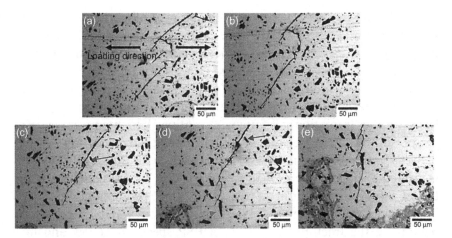

Figure 3.23 Change in crack growth path as a function of fatigue life: (a) after 40,000 cycles, (b) after 60,000 cycles, (c) after 80,000 cycles, (d) after 100,000 cycles, and (e) after 120,000 cycles. Crack that was growing at 45° to tensile axis is now almost at 90° to the loading direction. Si particles adjacent to the crack have been marked for easier identification of the crack extension event [17].

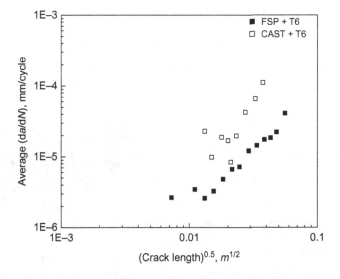

Figure 3.24 Crack growth rate curve, cast + T6 versus FSP + T6 [17]. Dashed line indicates transition to Paris regime.

the cast + T6 condition as compared to the FSP + T6 condition. The Paris exponent m is ~4 in the cast + T6 condition, while in the FSP + T6 condition, m was found to be ~2. Therefore, a higher crack growth rate together with a higher Paris exponent resulted in the lower fatigue life observed in the cast + T6 condition. Porosity present in a cast specimen acts as notches, and therefore results in a higher crack growth rate and lower fatigue life. Numerous instances of literature data show that casting defects can act as notches during fatigue failure [18,26]. Figure 3.24 further indicates that a very short Paris regime is operative in the case of the cast + T6 specimen, i.e., on the right-hand side of the dashed line. In general, the stage II or Paris regime of FCG leads to the formation of fatigue "striations" in many engineering alloys. In the Paris regime of fatigue crack advance, the spacing between adjacent striations correlates well with the experimentally measured average crack growth rate per cycle [18]. A significant variation in striation spacing in as-cast versus as-FSPed condition is noted in the work by Jana et al. [17]. Figures 3.25a and b show certain regions on the fracture surface of an FSP specimen at low and high magnifications. Characteristic striations with an average spacing of 0.5 μm are clearly visible in Figure 3.25b. On the other hand, the cast specimen fracture surface shows the presence of much coarser striations with spacing of ~1 μm (Figures 3.25d and e). Regions where final fracture took place also differ depending on the specimen type. Dimpled rupture morphology is prominent in the FSP specimen, while cleavage fracture is associated with the as-cast specimen (Figure 3.25c and f). This distinction is a result of the beneficial effect of FSP on the ductility of the alloy. As mentioned earlier, fatigue failure in castings

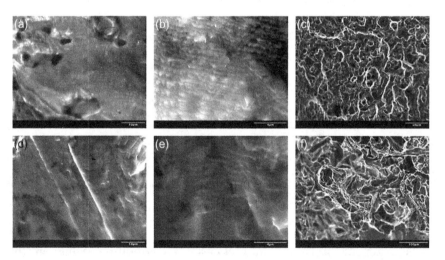

Figure 3.25 Comparison of SEM fractographs, FSP + T6 (a–c) versus cast + T6 (d–f): (b) shows fine striations in FSP + T6 condition, whereas coarse striations are observed in the cast + T6 condition, shown in (e); dimpled rupture morphology is evident in the FSP + T6 sample shown in (c), while the cleavage mode of fracture is visible in the cast + T6 sample, as shown in (f) [17].

(especially in Al-Si alloys) is dominated by FCGR because of the presence of casting defects, second-phase particles, etc. A poor fatigue life is noted in as-cast condition due to a notch effect shown by the defects. Additionally, the crack growth rate is also faster in the as-cast condition (at least 5×, as noted by Jana et al. [17]). Higher crack growth rate in the as-cast condition is associated with less meandering or tortuosity of fatigue cracks. As FSP removes defects and refines second-phase particles, the notch effect is removed, while highly tortuous FCG is observed, ultimately resulting in much better fatigue performance.

SHORT CRACK BEHAVIOR: ROLE OF MICROSTRUCTURE

An interesting aspect of FCG is its oscillatory behavior during the first few hundred cycles of fatigue testing (Figure 3.26). Such crack growth behavior is attributed to the "short" nature of the initiating fatigue flaws. Short cracks can be broadly classified as [18] (i) mechanically small, when the size of the crack is comparable to the plastic zone size, (ii) microstructurally small, when the crack size is comparable to the scale of microstructural characteristic dimensions (MCD), (iii) physically small, when the crack size is significantly larger than MCD or plastic zone size but has a length of ~ 1−2 mm, and (iv) chemically small, when the crack size is nominally amenable to linear elastic

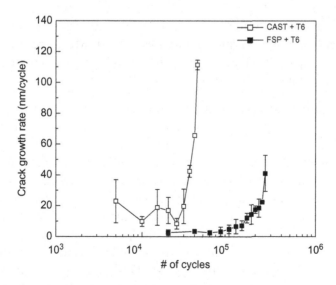

Figure 3.26 Oscillatory fatigue crack growth rate behavior, F357 alloy [17].

fracture mechanics (LEFM) analyses but shows apparent anomalies in propagation rate due to the effect of environmental stress corrosion fatigue on crack dimensions. It is well documented that short cracks undergo transient accelerated growth followed by retardations, resulting in oscillatory behavior below a critical crack size; however, beyond this critical size, crack growth rate data merges with that for long cracks [18,27−30]. In general, the retardation occurs whenever the crack tip encounters an obstacle in its path, which can be a grain boundary or a particle depending on the situation. The so-called anomaly related to the growth of short fatigue cracks lies in this retardation. According to LEFM, an increase in crack length should result in larger ΔK for a fixed far-field cyclic stress; therefore, a higher crack growth rate is expected. This is obviously not the case below a critical crack size, thus creating the problem of "short cracks."

Critical crack size, a_0, has been demonstrated to be related to the intrinsic threshold stress intensity factor range, ΔK_0, and smooth-bar fatigue limit, σ_e, by the following equation [18]:

$$a_0 = \frac{1}{\pi}\left(\frac{\Delta K_0}{\sigma_e}\right)^2 \qquad (3.8)$$

Efforts have been made to link a_0 with MCD in a vast spectrum of materials. Taylor [31] proposed a_0 to be $10d$, where d is the average grain size or another relevant microstructural distance. This distance can be mean-free path for a microstructure where plasticity is governed by barriers other than grain boundaries [31,32]. More recently, Lados et al. [33−35] asserted that the critical length of a microstructurally small crack should be between $5\times$ and $10\times$ MCD.

In the study by Jana et al. [17], oscillatory crack growth behavior was seen up to crack lengths of 450 μm for the as-cast condition, whereas in the as-FSPed condition oscillatory behavior stopped at 180 μm, as shown in Figure 3.27. For a cast alloy, secondary dendrite arm spacing (SDAS) is the critical MCD as it corresponds to successive Si-particle-dominated regions or mean-free path. Jana et al. [17] reported an SDAS of ∼100 μm in the studied F357 alloy. In the FSP condition, the MCD corresponds to interparticle spacing, $d_{Si\text{-}Si}$, which was measured to be ∼17 μm. Therefore, the critical transition crack length was between $5\times$ and $10\times$ MCD. The oscillatory behavior is due to the interaction of a short fatigue crack with a barrier, which

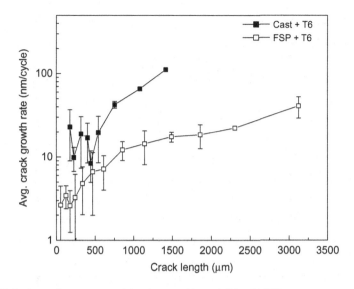

Figure 3.27 Crack growth rate versus crack length, cast + T6 versus FSP + T6 [17].

would be interdendritic regions in a cast condition and individual Si particles in as-FSPed condition.

FRACTURE TOUGHNESS

Fracture toughness of FSP microstructure in an A356 alloy has been determined by Sharma et al. [36]. Compact tension (CT) specimens were prepared and tested according to the ASTM E647 standard for measuring FCGRs. Crack growth tests were conducted at $R = 0.1$ and 0.5 at a sinusoidal frequency of 8 Hz. As the dimensions of CT specimens did not meet the plane strain fracture toughness criteria (ASTM E399), the upper limit of the crack driving force was designated "pseudo" fracture toughness. FCGR curves for different heat treatment conditions are shown in Figure 3.28. Improved FCG characteristics in the FSP condition is confirmed by Sharma et al. [36]. The FCG threshold (ΔK_{th}) is significantly higher for FSP samples than for cast samples. Moreover, crack growth rates in the cast condition are also higher. As load ratio is increased, a drop in ΔK_{th} occurs, suggesting the importance of various operative crack closure mechanisms. Results of the conducted fracture toughness test at $R = 0.1$ are summarized in Table 3.13. Much higher fracture toughness in FSP condition can be noticed, which is attributed to a more ductile Al matrix.

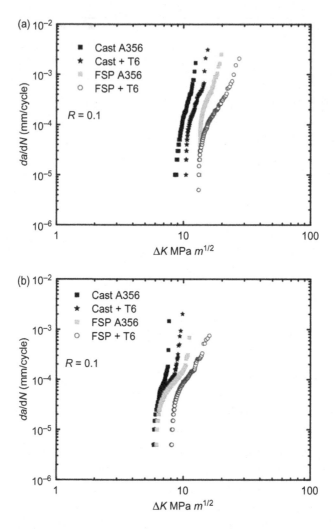

Figure 3.28 Crack growth rates in A356 under various processing conditions: (a) R = 0.1 and (b) R = 0.5 [36].

Table 3.13 Summary of Fracture Toughness Test Data in A356 alloy, $R = 0.1$ [36]				
Material	A, mm/cycle $(MPa\sqrt{m})^{-1}$	m	Yield strength (MPa)	K_Q (MPa $\sqrt{m}$)
Cast	2.29×10^{-13}	8.74	132 ± 3	14.6 ± 2
Cast + T6	9.33×10^{-8}	3.10	225 ± 20	15.8 ± 4
FSP	2.1×10^{-12}	6.84	168 ± 9	19.5 ± 1
FSP + T6	4.84×10^{-10}	4.59	288 ± 9	24.4 ± 1

PROBABILISTIC FATIGUE LIFE PREDICTION

It is well established that pores in a casting are the most detrimental defect in terms of fatigue life. Therefore, most of the currently available probabilistic prediction models compute fatigue life from the pore size distribution alone. However, as pore size decreases, fatigue life increases. Below a critical pore size, fatigue crack initiation through other defects, e.g., second-phase particles, becomes operative. Crack initiation through PSB formation can be another mechanism. Therefore, a more complete picture for fatigue life prediction could be achieved by combining the various defects, e.g., pores, second-phase particles, and grains.

Recently, Kapoor et al. [37] developed a probabilistic fatigue life prediction model, which is based on statistical distribution of pores, intermetallic particles, and grains. This model computes the probability of small crack initiation based on the probability of finding combinations of defects and grains on the surface. With the main assumption being that fatigue crack nucleation takes place on the specimen surface, the following combinations of defects for crack initiation were addressed:

- crack initiation from a pore—grain combination
- crack initiation from a particle—grain combination
- crack initiation from a grain.

A summary of relevant equations developed for each combination is presented in the same order in Table 3.14, while the overall algorithm of the model is shown in Figure 3.29. For experimental

Table 3.14 Microstructural Features Initiating Cracks, the Number of Cycles for Crack Initiation, and the Probability of Occurrence on the Surface [37]

Initiating Features	Crack Initiation Cycles	Probability of Occurrence
Pore of size D_i Grain of size L_j	$= \dfrac{2M^2 GW_s}{\pi(\bar{\sigma}_{a,loc} - \sigma_e)^2}\dfrac{1}{L_j}$ $\bar{\sigma}_{a,loc} = \dfrac{1}{L}\displaystyle\int_0^L \sigma_{a,loc}\,dx$ $= \sigma_a\left\{\dfrac{D_i k_t}{4L_j}\left[\left(1+\dfrac{8L_i}{D_i}\right)^{1/2}-1\right]+1\right\}$	$= \left[1-\left(1-\dfrac{SD_i}{V}\right)^{n_p(D_i)}\right]\cdot f_g(\ln L_j)\,\delta\ln L$
Particle of size ξ_i Grain of size L_j	$= \dfrac{1}{2}\left(\dfrac{G'+G}{G'}\right)\dfrac{M^2 GW_I}{(\sigma_a - \sigma_e)^2}\dfrac{1}{\xi_j}$	$= \left[1-\left(1-\dfrac{S\xi_i}{V}\right)^{n_{IM}(\xi_i)}\right]\cdot f_g(\ln L_j)\,\delta\ln L$
Grain of size L_i	$= \dfrac{2M^2 GW_s}{\pi(\sigma_a - \sigma_e)^2}\dfrac{1}{L_i}$	$= f_g(\ln L_i)\,\delta\ln L$

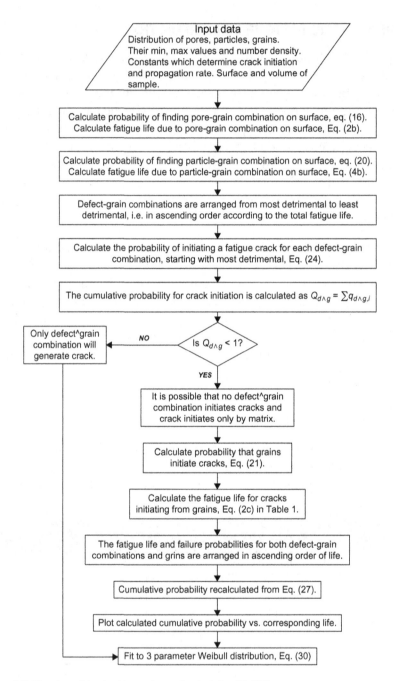

Figure 3.29 Flow chart of the algorithm used to predict the fatigue life [37].

validation of the model, fatigue testing of A206 alloy (Al-4.7Cu-0.4Mn-0.3Mg-0.2Ti, wt%) in as-cast and in as-FSPed conditions was carried out. The microstructure of A206 consists of primary alpha phase with needle-shaped Cu_2FeAl_7 intermetallic particles in the inter-dendritic regions and a significant volume fraction of porosity. FSP was carried out on A206 cast + T4 plate with a conical tool-steel tool using a rotation rate of 1000 rpm and a traverse speed of 2.54 mm/s. Predicted crack initiation life and total fatigue life, together with experimental data, are shown in Figures 3.30 and 3.31. The developed model provides a fair match with the experimental data.

Because the proposed model predicts the fatigue life before and after FSP of A206 fairly well, it could be used to predict life with any desired microstructure. As an example, if there were two sets of second-phase particles with different levels of refinement from FSP parameter optimization in a cast Al alloy, fatigue response would differ significantly. This is shown in Figure 3.32. A decrease of half in particle size distribution would approximately double the number density of particles. Keeping the same grain size distribution, the effect of particle refinement on fatigue life enhancement is noticeable.

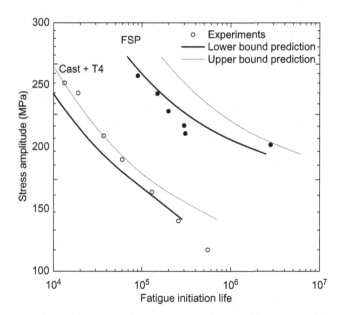

Figure 3.30 A comparison between the experimental and predicted stress amplitude versus fatigue crack initiation life for both cast + T4 and FSP conditions. The lower and upper bounds of the prediction are shown on the plot [37].

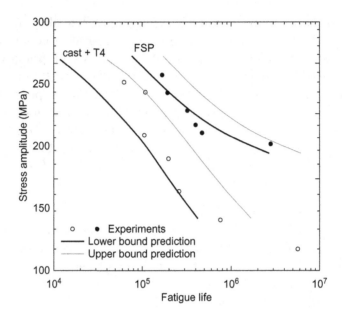

Figure 3.31 A comparison between the experimental and predicted stress amplitude versus fatigue life for both cast + T4 and FSP conditions. The lower and upper bounds of the prediction are shown on the plot [37].

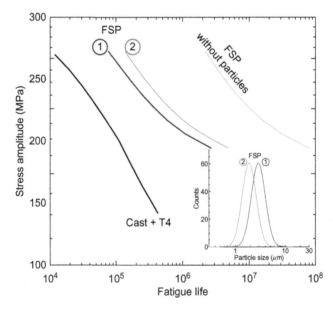

Figure 3.32 Stress amplitude versus lower bound of computed fatigue life showing cast + T4 condition (black) and FSP conditions from the actual defect distribution labeled 1 (blue), the distribution of smaller particles labeled 2 (gray), and a condition without particles (green). The inset shows the corresponding particle distributions marked 1 and 2 [37].(For interpretation of the references to color in this figure legend, the reader is referred to the web version of this book.)

Formation of fine grain structure after FSP without any intermetallic particles would be another desirable microstructure (Figure 3.32); the lower bound of the stress-amplitude–fatigue-life curve shifts to high numbers of cycles to fatigue failure. This offers a potential improvement in fatigue life if, along with the elimination of pores, the intermetallic particles could be eliminated.

REFERENCES

[1] Ma ZY, Sharma SR, Mishra RS. Metall Mater Trans A 2006;37A:3323.

[2] Jana S, Mishra RS, Baumann JA, Grant GJ. Metall Mater Trans A 2010;41A:2507–21.

[3] Nakata K, Kim YG, Fujii H, Tsumura T, Komazaki T. Mater Sci Eng A 2006;437:274–80.

[4] Kapoor R, Kandasamy K, Mishra RS, Baumann JA, Grant G. Mater Sci Eng A 2013;561:159–66.

[5] Sun N, Apelian D. JOM November 2011;44.

[6] Feng AH, Ma ZY. Scripta Mater 2007;56:397–400.

[7] Feng AH, Xiao BL, Ma ZY, Chen RS. Metall Mater Trans A 2009;40A:2447.

[8] Santella M, Frederick A, Degen C, Pan T-Y. JOM May 2006;56–61.

[9] Tsujikawa M, Chung SW, Tanaka M, Takigawa Y, Oki S, Higashi K. Mater Trans 2005;46 (12):3081–4.

[10] Freeney TA, Mishra RS, Grant GJ, Verma R. Friction stir welding and processing IV. Warrendale, PA: TMS; 2007.

[11] Freeney TA, Mishra RS. Metall Mater Trans A 2010;41A:73.

[12] Xiao BL, Yang Q, Yang J, Wang WG, Xie GM, Ma ZY. J Alloys Comp 2011;509:2879–84.

[13] Yang Q, Xiao BL, Ma ZY. Metall Mater Trans A 2012;43A:2094.

[14] Pilchak AL, Williams JC. Metall Mater Trans A 2011;42A:1630.

[15] Wang QG, Davidson CJ, Griffiths JR, Crepeau PN. Metall Mater Trans B 2006;37B:887–95.

[16] Sharma SR, Ma ZY, Mishra RS. Scripta Mater 2004;51:237–41.

[17] Jana S, Mishra RS, Baumann JB, Grant G. Acta Mater 2010;58:989–1003.

[18] Suresh S. Fatigue of materials. 2nd ed. Cambridge, MA: Cambridge University Press; 1998.

[19] Couper MJ, Nesson AE, Griffiths JR. Fatigue Fract Eng Mater Struct 1990;13:213–27.

[20] Wang QG, Apelian D, Lados DA, Light J. Metals 2001;1:85–97.

[21] Murakami Y, Endo M. The behavior of short fatigue cracks. London: Mechanical Engineering Publication; 1986. p. 275–93.

[22] Murakami Y, Endo M. Eng Fract Mech 1983;17:1–15.

[23] Lee FT, Major JF, Samuel FH. Metall Mater Trans A 1995;26A:1553–70.

[24] Gall K, Yang N, Horstemeyer M, McDowell DL, Fan J. Metall Mater Trans A 1999;30A:3079–88.

[25] Broek D. In: Pratt PL, editor. Fracture. London: Chapman and Hall; 1969. p. 734.

[26] Taylor D, Knott JF. Met Technol 1982;9:221–8.

[27] Miller KJ, de los Rios ER, editors. The behavior of short fatigue cracks. London: Mechanical Engineering Publications; 1986.

[28] Ritchie RO, Lankford J, editors. Small fatigue cracks. Warrendale, PA: TMS; 1986.

[29] Kitagawa H, Tanaka T, editors. Fatigue 90. Birmingham: Materials and Components Engineering Publication; 1990.

[30] Shyam A, Allison JE, Jones JW. Acta Mater 2005;53:1499–509.

[31] Taylor D. Fatigue thresholds. London: Butterworth & Co. (Publishers); 1989.

[32] Yoder GR, Cooley LA, Crooker TW. Metall Trans A 1977;8A:1737–43.

[33] Lados DA, Apelian D, Donald JK. Acta Mater 2006;54:1475–86.

[34] Lados DA, Apelian D, Jones PE, Major JF. Mater Sci Eng A 2007;468–470:237–45.

[35] Lados DA, Apelian D. Eng Fract Mech 2008;75:821–32.

[36] Sharma SR, Mishra RS. Scripta Mater 2008;59:395–8.

[37] Kapoor R, Sree Hari Rao V, Mishra RS, Baumann JA, Grant G. Acta Mater 2011;59:3447–62.

CHAPTER 4

Friction Stir Processing: A Potent Property Enhancement Tool Viable for Industry

INDUSTRIAL IMPLEMENTATION: POSSIBLE WAYS OF INTEGRATING FSP

Friction stir processing (FSP) has found commercial application in several niche products (microelectronics, cutting blades, vacuum system hardware), but high-volume applications have yet to surface. Several industries have recognized the potential and are actively researching opportunities to use FSP to improve product performance and efficiency in automotive, aerospace, heavy vehicles, consumer electronics, power transmission, and applications in the defense sector. Only a small number of these have been reported in the open literature. Internal research groups within manufacturing companies explore new technologies, often in collaboration with Universities, National Labs, or Contract Research entities, under nondisclosure environments to protect any early advantage that the new technology might provide in a competitive marketplace. As a result, it is often difficult to assess the technical readiness of a new technology until a product is revealed; at which point the technical readiness is quite high! Except for the niche commercial products, it is probably fair to put FSP at a Technology Readiness Level (TRL) of 4–5. Laboratory demonstrations of performance enhancement through FSP have been shown at full scale in relevant environments, but few have been demonstrated at the prototype part level integrated into subsystems (TRL6). To illustrate the readiness level, a few examples of some applications and current FSP research projects are described below.

This section gives a snapshot of some of the applications that are being investigated and some of the potential benefits FSP could provide to increased product performance and efficiency. FSP opportunities can be broadly grouped into three classes: (i) Applications where the outcome is an improved near-surface bulk property of the casting, (modulus, strength, resistance to fatigue failure, thermophysical property); (ii) applications

Friction Stir Casting Modification for Enhanced Structural Efficiency. DOI: http://dx.doi.org/10.1016/B978-0-12-803359-3.00004-2

where the improvement involves how the casting interacts with the environment (surface improvements—wear resistance, surface microstructure, or finish for later surface treatment); and (iii) applications where FSP is used to repair a high-value casting after an in-service flaw has been detected.

Bulk Property Improvement

FSP has been investigated for several applications in internal combustion engines where durability affects product lifetime. In many cases, the durability of the cast part is dependent on the presence of a high-performance microstructure in only a small area of the part. To satisfy the needs of this small area, often the entire part is required to have the high-performance microstructure. Examples include Sr-modified bulk alloys or costly casting procedures to insure an appropriate secondary dendrite arm spacing exists in a specific area for fatigue performance. FSP's ability to create high-performance microstructures only in local regions may allow for lower cost materials or processes in the bulk of the part. Examples under investigation currently are FSP of diesel pistons for improved bowl rim fatigue, FSP of cylinder head castings for improved fatigue performance in the valve bridge area, FSP of block castings for local strength and fatigue improvement, and FSP of cast crankshafts to improve the fatigue performance at local stress concentrations such as oil holes and fillets. Figures 4.1 and 4.2 illustrate these concepts.

Figure 4.1 shows a concept for using FSP to improve the fatigue performance of diesel pistons. Driving this application is the need to go to higher efficiency combustion strategies for increased energy efficiency. New combustion strategies require significantly increased peak cylinder pressure leading to an environment where fatigue cracking of the bowl rim area on the piston top creates unacceptable product lifetime.

Figure 4.1 Diesel piston top showing fatigue cracks that lead to failure of the bowl rim.

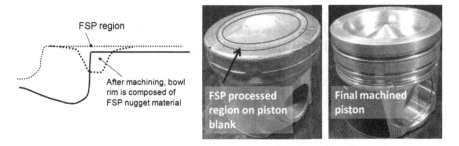

Figure 4.2 FSP applied to an aluminum diesel piston top to improve the fatigue performance of the bowl rim.

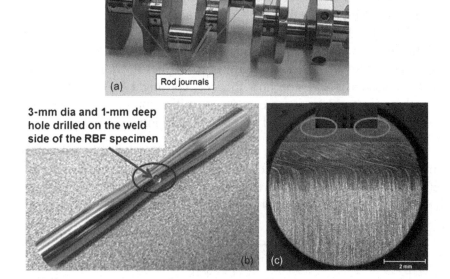

Figure 4.3 Typical crankshaft after finish machining showing oil hole locations (a), RBF specimen showing blind hole to simulate stress concentration of oil hole (b), cross-section of an FSP process specimen showing blind hole edges located in the FSP microstructure (c).

Very costly changes in piston material are one solution. This study demonstrated that if FSP could be done on the piston blank prior to final machining such that the stir zone material formed the exposed bowl rim, then fatigue failure could be significantly suppressed.

Figure 4.3 illustrates a study on using FSP to improve the fatigue performance of a crankshaft. An internal combustion engine uses a crankshaft that is internally lubricated by a flow of oil through passages that breach the surface at each bearing journal. These holes on the surface where the oil passages emerge can have significant stress concentrations and locations of early fatigue failure. The concept of this study is to friction stir process the area on the casting blank that will later be drilled for the oil passage. In this way, the edges of the hole and the stress concentrations associated with those edges would be located in the fine-grained, fatigue-resistant microstructure left over from the FSP step.

As a crankshaft analog, rotary beam fatigue testing was done on cylindrical samples of base metal, base metal with short blind holes drilled in it, and base metal with FSP regions created, then later drilled with a similar blind hole. Testing revealed that holes drilled in base metal produced a 4× drop in fatigue life, but when the holes were drilled in FSP-processed regions, the samples recovered the full base metal fatigue life (Figure 4.4). In this case, the FSP microstructure significantly suppressed the fatigue initiation and/or growth of cracks associated with the stress concentration. It does not take a great deal of imagination to envision how this concept might be broadly applied to other structural environments where mitigation of stress concentrations is driving design toward a high-cost solution.

Another factor creating a pull for FSP is the desire in automotive and on-road heavy vehicle to reduce vehicle mass to increase energy efficiency in use and to decrease the amount of embedded energy in the manufacturing of the raw material. FSP can be an effective tool to

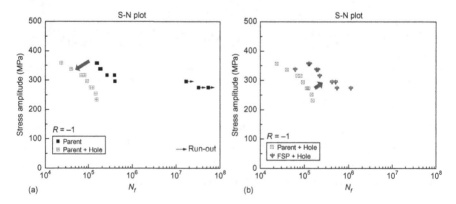

Figure 4.4 When a hole is drilled in the parent material, the fatigue life drops by a 4× (a). However if the hole is drilled in FSP material, the full life is recovered (b).

selectively improve the performance of a casting only where the performance is required. In many cases, castings are required to have inefficient heavy wall thickness to satisfy a structural requirement that is generated in only a small portion of the casting. If the casting can remain with down-gaged wall thickness, but, through FSP, show improved properties only where needed, mass savings can occur, and both front-end and end-user energy efficiencies can be gained. An example from the manufacturing of aluminum automotive wheels is included in the section on Role of Numerical Tools.

Another application for FSP currently being investigated involves the need for a local high-performance microstructure to improve confidence around a secondary machining operation. There are numerous examples of secondary operations such as finishing, drilling, threading, etc. in which the local region to be machined is required to have a characteristic such as fine grain size, microstructural homogeneity, or lack of porosity. Also, in many cases, drilling and tapping operations are performed on castings in areas of the casting that are difficult to cast (thin flange features, last to fill, etc.). FSP has been investigated as a way to insure a dense homogeneous microstructure that exists prior to drilling and tapping.

As an alternative to drilling, often the casting is designed with cast holes that are later tapped directly for fasteners. These holes formed by die inserts or die pins can have porosity and blistering of the cast material adjacent to the cold pin. FSP has been investigated as a way of producing dependable microstructures in these environments as well.

Another application of FSP investigated for improved near-surface microstructure for secondary machining operations comes from the vacuum products industry, but can also be applied generally to sealing surfaces. FSP has been found to be valuable in creating a homogeneous microstructure important for knife edge or o-ring seal grooves on a sealing surface flange. As can be imagined, improved seal performance can have implication across industries (internal combustion head gasket, high pressure or ultra-low vacuum piping and pressure vessel, etc.).

More exotic applications for FSP have also been investigated to create microstructures and textures to improve magnetic response in magnets or the magnetic field response in chambers containing ion beams.

FSP has also been investigated to modify the thermophysical properties of castings. Several researchers have used FSP to embed ceramic particulate or carbon nanotubes to modify the near-surface thermal conductivity or electrical properties. A wide range of applications could benefit from customized or designed thermal properties (heat sinks, heat pipes, thermal barrier coatings etc.).

One distinct advantage that FSP has over other secondary operations that may be considered for local property improvement is the relative ease that the process can be embedded in a manufacturing line. FSP is fundamentally a machining process achieved by equipment that in principle is no different than an NC milling center. In light alloys and in shallow penetration environments, FSP is easily accomplished with existing milling machines. It is envisioned that the application of FSP to a part could be easily integrated into a fabrication line without the need to de-fixture the part. In the case of a milling center with an automated tool changer, an FSP tool could be used as easily and as interchangeably as a cutting tool. The process, after up-front "weld" process efforts are completed, is completely specified numerically and does not require the intervention of an expert operator.

Environmental Interaction

A second broad category where FSP has been investigated by government, industry, and academia is where there is a need to improve the performance of the actual surface of the casting for a particular purpose. Most investigations of this type have focused on improved wear resistance. An example of this work is studies done on improving the wear performance of cast iron for application in brake rotors. In this work, FSP was used to physically stir in a TiB_2 powder into the surface of the cast iron. This powder reacted in a solid state to form a very fine-grained and homogeneously distributed TiC reaction product. This FSP-processed material was tested by ATSM G35 and found to show roughly twice the wear resistance of the baseline brake rotor material (Figure 4.5).

Another application where surface properties are improved by FSP is in cases where the surface needs to have a specific characteristic for a later coating application. Several applications have emerged and are under investigation by government, academia, and private industry on using FSP to prepare a surface for anodization. Anodization whether

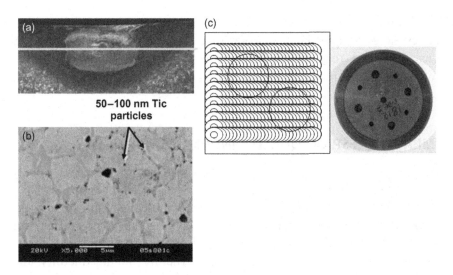

Figure 4.5 FSP region in cast iron (a). Microstructure showing fine-grained TiC reaction products (b). Raster pattern of FSP applied to surface and extracted rotor after testing (c).

for appearance or for some function such as environmental or physical degradation resistance is made more robust if the substrate has particular characteristics related to microstructure, homogeneity, and texture. FSP has been investigated as a method to produce appropriate surface microstructures for coatings and anodization in both automotive and the consumer electronics industry.

Potential as Repair Technique

A third application area where FSP may see a pull into industrial application is use as a repair process to repair surface cavitation or structural damage or cracking. The need for repair of a casting occurs at two distinctly different stages. The first one is at the initial stage of manufacturing of components. The solidification shrinkages and thermal gradients can lead to cracking, particularly in complex castings. Because of the high value of complex castings, it is desirable to repair rather than discard. Traditionally, castings are repaired by fusion welding techniques. So, microstructurally it remains limited and the region of repair still remains vulnerable. The obvious advantage of using FSP to repair such regions would be the enhancement of properties, which will make these regions better than the remaining cast regions. A challenge for this concept is that the repair surface needs to be accessible to the friction stir head of the machine and the local cross-section must be able to support the process forces.

The second type of repair need is after in-service cracking. Again, it is desirable to repair large castings that are costly and put back in service. The fact that the FSP repair would lead to better properties means that the repaired region will have better properties than the parent regions.

IMPACT OF FSP ON QUALITY INDEX OF CASTINGS

As highlighted in Chapter 3, the ductility of casting alloys is usually low, and changes to the casting process as well as changes in the chemical composition and/or heat treatment aimed at improving the strength or other properties can be used to enhance prroperties for structural applications. It is thus important to simultaneously assess the effect of any changes to the microstructure on the ductility and strength of the material. Hence, castings are evaluated using strength–ductility diagrams known as quality index charts [1].

High tensile strength and good tensile ductility are the most desirable properties in structural design, and if the quality index chart is used to plot the experimentally determined tensile strength and tensile ductility for a particular alloy, the best "quality" material will be located near the upper right corner. Different materials or processing conditions can thus be assessed on the basis of their locus on the chart. As noted in the Introduction section, this is partly the logic behind the development of the quality index charts [2]. Drouzy and co-workers [1] defined quality index Q (in MN/m^2) as:

$$Q = \sigma_{UTS} + 150 \log E \qquad (4.1)$$

where σ_{UTS} is the ultimate tensile strength (MN/m^2) and E is the elongation (%). The number 150 is an empirically determined constant applicable to alloys in the A356 and A357 (Al-7Si-Mg) systems. Drouzy et al. [1] noted that normal aging of a particular batch of cast alloy shows a linear relationship between σ_{UTS} and $\log E$ with a slope close to 150. If steps are taken to improve the quality of the casting by, for instance, reducing porosity, then a new parallel line is formed at higher levels of strength and/or ductility [2,3]. In its most straightforward application, the Q-values allow for a comparison between different alloys, or between batches of samples of the same alloy. As stated previously, high Q-value lines are close to the upper right corner of quality index plots and indicate that the material has both high UTS and high ductility, i.e., its mechanical quality is high. In general, a Q-value above 400 MPa is considered very well for alloy A356 [4]. It is normally assumed that the Q-value measures the "quality" of the casting as determined by the content of Fe-rich intermetallics, the degree of modification or process-related parameters such as porosity, dross, and inclusions. Originally, the quality index was given no explicit physical meaning, but research done by Caceres [2,4] have presented analytical models that relate the quality index to the necking onset strain of the material using continuum mechanics. This analysis allows calculation of the quality index from the knowledge of the parameters of the deformation curve and it is suggested [2] that it can be used as a tool for alloy design. The analytical model described by Caceres [2] is based on the assumption that the deformation curves of the material can be described with a constitutive equation of the form,

$$\sigma = K\varepsilon^n \tag{4.2}$$

where σ is the true flow stress, K is the materials strength coefficient, ε is the true plastic strain, and n is the strain hardening exponent defined by,

$$n = \frac{\varepsilon}{\sigma}\frac{d\sigma}{d\varepsilon} \tag{4.3}$$

The engineering stress, P, and the true stress, σ, are related by $\sigma A_f = P A_0$, where A_0 and A_f are the cross-section areas in the initial and strained conditions, respectively. The strain has elastic and plastic components, ε_{el} and ε_{pl}, so that $\sigma = Pe^{(\varepsilon_{pl}+2\nu\varepsilon_{el})}$ and Eqn (4.2) can be rewritten as follows:

$$P = K\varepsilon_{pl}^n e^{-(\varepsilon_{pl}+2\nu\varepsilon_{el})} \cong KS^n e^{-S} \tag{4.4}$$

where ν is Poisson's ration and $\cong$ indicates that the correction due to elastic deformation as well as the difference between nominal and true strain have been ignored. The latter is a reasonable assumption for casting alloys due to their limited (less than 3%) tensile ductility. But as has been noted in Chapter 3, the ductility after FSP is quite high in a lot of cases, in that situation the difference between nominal and true strain can become significant. Equation (4.4) can be modified to generate the iso-quality lines by introducing the relative ductility parameter q, defined as the ratio between the strain hardening exponent and the elongation to fracture S_f.

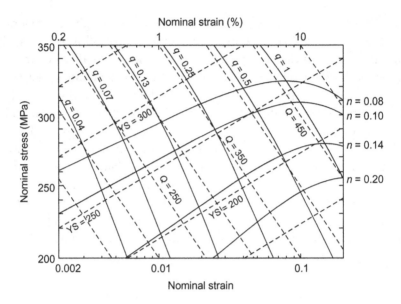

$$q = \frac{S_f}{n} \qquad (4.5)$$

Solving Eqn (4.4) for n and substituting in Eqn (4.5) gives

$$P = KS^{s/q}e^{-S} \qquad (4.6)$$

The lines generated by Eqn (4.6) meet the condition of constant relative ductility and are called the iso-q lines as shown in Figure 4.6 [2], each identified by a q value. The meaning of q is such that when $q = 1$ the sample reaches necking (most ductile samples) while $q < 1$ identifies progressively less-ductile samples. The correlation between iso-q lines and iso-Q lines shown in Figure 4.1 provides a straightforward physical meaning for the quality index in terms of relative ductility parameter. Also shown are the solid flow lines identified by the n-value.

The original quality index chart was developed for alloy A356 and thus its use for other materials conveys the implicit assumption that the parameters involved, particularly the slope, d, of the iso-Q lines in Eqn (4.1), do not depend on the material. Drouzy et al. [1] included an explicit warning on this regard in the original publication, and in fact, it has been shown experimentally that the slope and position of the

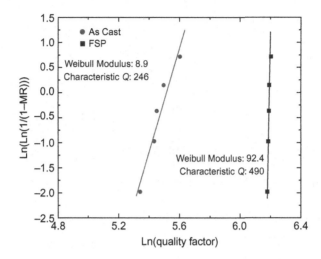

Figure 4.7 A plot of quality index showing significant in Weibull modulus after friction stir processing of A356 alloy [5].

iso-Q lines in the quality index chart change with both the chemical composition and temper in some alloys.

The entire discussion of the quality index is focused on the issue of reliability. Weibull modulus approach is often taken to discuss the reliability of materials. Figure 4.7 shows a quality index plot for A356 alloy using Eqn (4.1) [5]. The improvement in Weibull modulus after FSP is quite obvious and this suggests high reliability after FSP. This approach of reliability enhancement is analogous to the overall discussion of difference between cast component and forged component. The refined microstructure after FSP narrows the microstructural distribution and that leads to narrower distribution in property data.

ROLE OF NUMERICAL TOOLS: FEA OF STRUCTURES

Muppur et al. [6] carried out a case study involving finite-element-based fatigue life calculations for various A356 samples and components including an automotive wheel rim in cast and friction stir processed conditions. In this section, an illustrative example is included to show the effect of embedded FSP region on overall fatigue life performance of the wheel.

Figure 4.8 shows the solidification pattern in a wheel and the stress distribution. In the conventional design approach, the first step is to

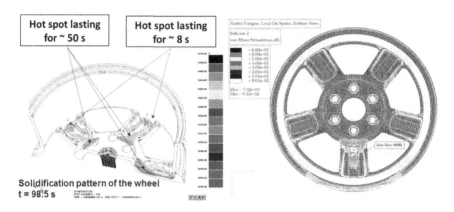

Figure 4.8 An example of an aluminum wheel showing the solidification pattern and the fatigue stress distribution.
Source: Courtesy Chuck Russo, Amcast, 2002.

do a stress analysis of the component. The cross-section of regions that would experience high stresses are increased. This leads to higher solidification time and lower cooling rate in those areas. This of course results in the worst cast microstructure in the regions that are most critical! For advanced castings, an alternative approach is to use "chill blocks" to manipulate the cooling rate. It does lead to finer dendritic microstructure but the overall cast microstructural features still there.

FSP offers an alternative to enhance the microstructure and properties in those localized region [7]. Muppur et al. [6] used the fatigue data from Sharma et al. [8] and the loading case methodology suggested by Stearns et al. [9] to simulate the impact of FSP on the enhancement of fatigue life of an automotive wheel. The service load for which the wheel was designed is 2480 KN. The service load was equally divided and applied to the curb-side bead region and the inner bead. The loads were uniformly distributed over an arc of 40 degrees. This was estimated to be the contact surface for a moving vehicle.

They used the stress-life approach because of its simplicity and a designer's interest in the total life of a component. Nodal values were used for fatigue calculations. Because the mesh was fine enough in all the cases, the nodal results took care of the load transfer in the critical regions. The fine mesh also ensured that averaging nodal results will retain the peak stress to a large extent. The maximum absolute principal stress tensor was chosen as the stress parameter because it was the driving factor for crack initiation and growth and thus can be assumed to be the primary influence on the total life of the component.

Two peak level loading cases are possible for the wheel rim: (i) when the spoke of the wheel takes the brunt of the load, and (ii) when the "window region" takes the brunt of the load. In either case, it is observed that the spoke-region gets highly strained. These areas were marked for FSP. Life contours of the cast wheel are shown in Figure 4.9a. It predicts minimum life in the high-stress regions of the spoke which was expected. These areas were marked for FSP and the properties of the nodes in these regions were changed to FSP-related mechanical properties. Only the nodes with peak stresses were selected. As a representation of the region, the depth and approximate volume of the region modified is calculated. Figure 4.9b shows the life

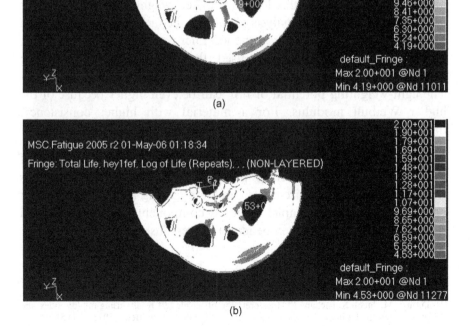

(a)

(b)

Figure 4.9 Log of life contours: (a) wheel rim in cast + T6 condition, and (b) wheel rim after FSP + T6 of critical regions.

Table 4.1 Impact of Selective FSP Regions on the Fatigue Life of Automotive Wheel Rim	
Condition	Life (Number of SAE Cycles)
Fully cast	15,488
Critical regions FSP	33,884
Each SAE fatigue loading cycle represents 1988 cycles with varying load amplitude.	

contours of the rim after FSP. Table 4.1 shows the life before and after FSP of the high-stress regions. Each SAE fatigue loading cycle represents 1988 cycles with varying load amplitude. The results show that the fatigue life doubles because of the selective embedded regions.

DESIGN CONSIDERATIONS

From Chapter 3 and the last section, it is clear that the embedded approach has potential to enhance the performance. There are two simple paths that can be taken with selective FSP approach. First, the designer can use this approach for light-weighting of components. As mentioned earlier, casting cross-sections are thickened in the regions of higher stress. Of course this leads to higher weight as well as impacts reliability. An economic analysis of this can be included by comparison of methods that can locally enhance microstructure and properties. For example, if chill blocks are used, then the cost of this approach and productivity aspects can be compared with the FSP approach. The light-weighting potential of FSP can be even higher because of the higher Weibull modulus. For a material with higher consistency, the safety factor used in design can be lowered.

The second approach can be to use FSP to extend the life of components. This approach is particularly applicable for situations where performance requirements eliminate casting as a manufacturing process choice. For example, a forged component can be substituted by a cast component with embedded FSP regions.

REFERENCES

[1] Drouzy M, Jacob S, Richard M. Interpretation of tensile test results using quality index and probable elastic limit—applications to cast Al-Si-Mg alloys. Rev Metall 1978;75(1):51−9.

[2] Caceres CH. Microstructure design and heat treatment selection for casting alloys using the quality index. J Mater Eng Perform 2000;9(2):215−21.

[3] Din T, et al. High strength aerospace casting alloys: quality factor assessment. Mater Sci Technol 1996;12:269−73.

[4] Caceres CH. A phenomenological approach to the quality index of Al-Si-Mg casting alloys. Int J Cast Metal Res 2000;12:367−75.

[5] Mishra RS, De PS, Kumar N. Friction stir welding and processing: science and engineering. International Publishing Switzerland: Springer; 2014. p. 284.

[6] Muppur SP, Mishra RS, Krishnamurthy K. Finite element based fatigue analysis of friction stir processed A356 components, unpublished research; 2006.

[7] Mishra RS, Sharma SR, Mahoney MW. Embedding a wrought microstructure in a cast component: a novel application of friction stir processing. Paper 05-086. Schaumburg, IL: Transactions of the American Foundry Society; 2005, vol. 113: p. 139−43.

[8] Sharma SR, Ma ZY, Mishra RS. Effect of friction stir processing on fatigue behaviour of A356 alloy. Scripta Mater 2004;51:237−41.

[9] Stearns J, Srivatsan TS, Prakash A, Lam PC. Modeling the mechanical response of an aluminium alloy automotive rim. Mater Sci Eng A 2004;368:262−8.

CHAPTER 5

Summary and Future Outlook

To wrap up this volume, the top-level aspects are captured along with some final thoughts of how to move the field of friction stir casting modification forward. Some of the key impediments are captured again.

The first aspect is related to the level of microstructural modification. The casting porosity is completely eliminated during friction stir processing (FSP) and it offers designers and practitioners different alternatives about cost involved in controlling porosity. For example, use of chill blocks can be eliminated for castings by incorporating FSP in localized areas. Although the chill blocks reduce the pore size by enhanced cooling rate, FSP eliminates it completely by consolidating the material. Additionally, other casting defects like folded oxide layers, gas porosity, etc. are also eliminated during FSP or their size is significantly refined. The next level of microstructural feature is the second-phase particles. Coarsest second-phase particles normally are the constituent particles, typically Fe and Si containing in aluminum alloys. The conventional way of overcoming the size issue of these particles is chemical modification. For example, Sr modification in Al-Si casting alloys leads to refinement of eutectic Si particles. Again, the FSP provides an alternative to the chemical approach. The refinement of second-phase particles during FSP is mechanical in nature. This is where the limit of this process needs to be acknowledged. Because of the strain gradient in the stir zone, some level of microstructural gradient should be expected. Similarly, because of elevated temperature during the material flow, the extent of shear stress that can be imposed on the second-phase particle is limited. So, the refinement beyond submicron size depends on the chemistry and shear strength of the second phase. The selection of process parameters is important and the principle is quite different from friction stir welding, where the objective may be to traverse as fast as possible. For friction stir casting modification, the advance per revolution needs to be controlled to achieve desired refinement and homogenization. Achieving homogenization during FSP can cut down on the extent of heat treatment required, thereby cutting heat treatment cost.

Friction Stir Casting Modification for Enhanced Structural Efficiency. DOI: http://dx.doi.org/10.1016/B978-0-12-803359-3.00005-4

An intriguing possibility of friction stir casting modification involves post FSP solution treatment. If the solution treatment step of the casting is moved to after FSP, i.e., cast $\rightarrow$ FSP $\rightarrow$ solution treatment, then an abnormal grain growth regime opens up to obtain very coarse grain size. This opens up the final microstructural domain to obtain fine grains or coarse grains. Certain high temperature applications where creep is important, such large grains may be desirable.

The impact on properties is directly related to final microstructural length scale and distribution. If one considers several design approaches, it is easy to scope the domain where friction stir casting modification can make major impact. Overall the argument for friction stir casting modification has to be based on the performance. Among various structural design approaches, let us consider (i) stiffness limiting design, (ii) strength limiting design, (iii) fatigue limiting design, (iv) toughness limiting design, and (v) creep limiting design. Of course, the stiffness is microstructure insensitive, and therefore no gains are expected from just microstructural modification. The increase in strength after friction stir casting modification is alloy dependent. The gains can be very significant for alloys that have significant property difference in cast and wrought conditions. For example, the strength of A206 aluminum alloy can go to the level of 2XXX aluminum alloys. This can be used effectively to lower the weight of castings by selective enhancement of strength by FSP of high-stress regions.

Fatigue properties are very sensitive to microstructural flaws like porosity and inhomogeneity. Friction stir microstructural modification can be used to enhance life and reliability. Coupling the experimental work with probabilistic modeling will allow maximum design flexibility by taking advantage of full potential. Several orders of improvement in fatigue life is possible with this approach. For components that are designed for fixed life cycle, weight savings can be a big driver.

Dependence of toughness on microstructure is complicated by the mechanics of stress concentration. The final outcome does depend on the specific alloy being considered. For example, while microstructural modification leads to enhancement in toughness and fatigue crack growth resistance in aluminum alloys, similar enhancement is not observed in magnesium alloys. So, it is important to quantify the intrinsic microstructural dependence of toughness in a specific alloy.

Several nonstructural applications like elimination of porosity on a sealing surface are possible, but this is also dependent on economic analysis. Availability of friction stir technology at the casting suppliers is going to be a key step in technology diffusion of this approach.

The overall impact of friction stir casting modification will also depend on the design approaches, which has to start at the original equipment manufacturers. Part of the motivation for adoption of new technology and concepts also comes from the demand for higher performance. Historically, high-performance systems have been driven by challenging needs or legislative requirements. For example, consider higher corporate average fuel economy (CAFE) standards for automotive sector. Higher CAFE standards led to choice of higher performance materials and manufacturing processes. In the absence of such demands, economics trumps performance for nonluxury automobiles. Aerospace sector lives on high performance but is generally conservative in adoption of new technology. Again, higher performance demanded by defense sector usually leads to innovative designs and selection of leading manufacturing processes. Designers are often the lynchpin for adoption of new technologies and friction stir casting modification is no different. It needs leading designers to design components that take advantage of "embedding forged microstructure in cast components."

Printed in the United States
By Bookmasters